Millions From the Mind

How to Turn Inventions (Yours or Someone Else's) Into Fortunes

Alan R. Tripp

amacom
American Management Association

Notice to Readers

This book is intended to provide interesting and useful information about the process of turning inventions into successful commercial products. Neither the author nor the publisher is in the business of offering legal, accounting, or financial counsel, and, although the contents of this book are, to the best of the author's knowledge, factual, the reader should make business decisions based upon his or her own judgment and appropriate professional advice.

Library of Congress Cataloging-in-Publication Data

Tripp, Alan R.
Millions from the mind : how to turn inventions (yours or someone else's) into fortunes / Alan R. Tripp.
p. cm.
Includes bibliographical references.
ISBN 0-8144-5025-3
1. New products—Management. 2. Inventions—Economic aspects. I. Title.
HF5415.153.T75 1992 91–43234
658.5'752—dc20 CIP

Printing number

10 9 8 7 6 5 4 3 2 1

For
Maggie
. . . to whom this book
would have been dedicated
even if she were not my wife.

Contents

But First, This Message . . .

This book is about inventions and money. Specifically, it's about technology-based new products that have made millions of dollars for inventors and for corporations.

In part, it's a "how to do it" book, providing maps and rules of the road based upon dozens of money-making innovations. But equally, this is a "how they did it" book, revealing what the millionaire winners did—and what they avoided doing—along the road to royalties.

On the pages that follow you will read about a man whose "simple" invention of a sheet metal screw brought him not less than a million dollars annually in royalties for fifteen consecutive years; about a toothbrush invention launched by a start-up company that acquired five years later for $110 million; and about an R&D man in a Fortune 500 company who fought off the engineers when they said his invention couldn't be manufactured, who sloughed off the market researchers who said the product's potential was too small, who helped work out a selling method when the company was ready to close down the test market—and who lived to see his invention reach $100 million in annual sales.

Other innovators discussed in this book made discoveries that earned only a million dollars or two. But they all created inventive solutions to the obstacles that stand between great ideas and great rewards. These inventors and businesspeople have in recent years proved that there is still opportunity in America to create wealth from the fruit of the mind.

But it is not easy. Inventions that create valuable new products and new jobs and new wealth do not spring to life full-blown as Athena was reported to have sprung from the head of Zeus.

To the contrary, after the conception of an invention, the hard work really begins. I call this *the business side of invention.* It encompasses a skein of skills that range from protecting the innovation to proving manufacturing feasibility to evaluating the market potential to finding the distribution channel to defending against copycats and much more—including raising enough money to go the distance.

In addition to optimism and sheer doggedness, converting an invention from a great idea to a money machine often requires the imprint of the entrepreneur. Esprit, elan, and invincible faith are always valuable. But they are not enough. The people who make money from inventions either understand or find allies who understand what turns patent examiners on and off, what triggers approval from financial people, how corporate executives operate, how technical and market research can be used or abused, when a new product should be launched by a start-up company, and when it is wiser to license it. And much more.

That's why the focus here is on the practical moves made by men and women who have actually made *millions from the mind.* I've combined them with my personal experiences over two decades in converting independently developed inventions into commercial products.

Although you will find rules, tools, and practical guidance here, please note this is not a textbook, but rather a thinking person's guide to turning creativity into cash. I have aimed to immerse your mind in the strategies and tactics that most often work, into the educational details of how the winners won—from there, discerning readers can apply the insights to their own inventions as they see fit.

Inevitably, one or more of the successes recounted here will have faltered or crashed or been surpassed. But I have tried to select cases where the operative principles hold true even when events outrun the stories.

So, if you are yourself an inventor, a discoverer, or a creative idea person inside or outside of a big corporation, I believe this book will give you not only the vicarious enjoyment of reading about the successes of others but direct insights into how to break through the barriers to commercializing invention.

If you are a corporate executive—in management, marketing, or research—I expect you to find ways to get the best results from the creative minds both within and outside your own company—and to manage the risks of new inventions without upsetting the entire board of directors.

If you are a financial person, investing your own money or the funds of others, I trust that you will find much here to sharpen your judgment as to what kind of inventions and what chain of circumstances offer the best assurance of ultimate success.

Historically, new products—specifically new products in which the benefit is based upon new technology, not just advertising or repositioning—have been the lifeblood of growth for corporations and for the United States as a whole. Western Electric, Singer Sewing Machine, and Curtis-Wright, to name a few, were founded by inventors who grew their companies and grew rich themselves.

More recently, SmithKline's Tagamet, Apple's Macintosh, Dupont's Lycra, and Gillette's Sensor are examples of innovative "point" products that have provided momentum for an entire company.

Yet today as we reach the end of a century, the majority of "new products" reaching the market are variations of old products: minor improvements, line extensions, new flavors, new colors, etc.

As every great inventor knows, true success does not come from imitating anybody. It comes from observing what works and what does not. It comes from observing unexplained phenomena and failed experiments, and from working on solution "B" when everyone else is still working on "A."

With the Japanese challenging us on one side and an economically united Europe rising on the other, it's time for Americans to pay more attention to invention. I hope that this book will be a useful tool for moving great ideas into production—and for making major money for those who do it.

Acknowledgments

Writing a book on how to make inventions make money required help from many sources, so many that I begin by assuring anyone I have omitted that the fault lies in my computer's memory and not in my heart.

I do warmly remember and thank the following:

For insights to legal points: Morton Amster of Amster, Rothstein & Ebenstein; Louis Heidelberger and Stephen Nagler of Reed Smith Shaw & McClay; Robert Frank of Choate, Hall & Stewart.

For leads and details on stories: Dr. William Davis of the University of California; Alistair Mann of Pfizer, Inc.; Jay Cohen of the American Society of Inventors; David Krantz, Australian architect; Gerald Kaminsky of Cowen & Co.; Peter Davis of SRI International; Irv Begelsdorf, science writer; Dr. Floyd Grolle of Stanford University; Alan Toffler of Ivanhoe Capital; Rom Cartwright of the oral hygiene division of Pfizer; Lisa Linden of Howard Rubenstein Associates.

For teaching me the business side of invention: Ray Cobb and Adrian Huns of the Boots Company in England; James Church, formerly of Marion Laboratories; Don Johnston, formerly of Schering-Plough; Robert Goldscheider of the International Licensing Network; Thomas Bogaard, a very inventive businessman.

For great courtesy in finding information for me: Sue Hagen and Nancy Arnold of the Newcomen Society; Deborah Cox of Colorocs Corp.; Joan Sesnick of Refac Technology; and some very professional librarians at the New York Public Library and the Villanova University Library.

For first aid in researching and editing: Vijay Kothare and Louise Stone.

For direct help and direction without which this book would still be an inventor's dream: Nancy Love, my friend and agent; Myles Thompson, Acquisitions Editor of AMACOM Books, who believed in this book from the beginning; Kate Pferdner, Managing Editor, and Richard Gatjens, Associate Editor of AMACOM Books; and, most decidedly, Eleanor Gates, my copy editor.

Part I

Money: The Creative Part of Invention

Chapter 1

Going for the Gold

Some wealth comes from the ground—oil, gas, copper, silver, gold, diamonds. You simply take it out of the earth and sell it. Other wealth comes from owning title to a piece of the earth's surface, a piece that is in the right location at the right point in time. Still other wealth comes from enjoying a monopoly or a quasi-monopoly position: The government grants you a right to lay down tracks and run a railroad, or a license to operate a broadcasting station, or a patent to manufacture and sell a cure for the common cold.

Spiritually at least, the source of wealth closest to the American way is The Big Idea: the invention, the discovery, the new product that springs from a pioneering mind to become an overnight success, bringing to its creator wealth beyond the dreams of avarice—or, *de minimis,* enough money to send the kids through college and to join a golf club.

A Tradition of Ingenuity

Well painted in the roseate reporting of American history books are such giants as Eli Whitney, the Boston lawyer who invented the cotton gin; Samuel Colt, who created a new level of mass production through highly accurate interchangeable parts; George Eastman, who made the first cameras with roll film; Alexander Graham Bell and Elisha Gray, who not only fought over which of them had first invented the telephone but found time for many other inventions, such as the gramophone and the predecessor of today's telefax machines; Thomas Alva Edison, who invented the light bulb and pioneered the concept of an "invention laboratory"; Wallace Hume Carothers, who worked for Du Pont and created nylon, and Roy J. Plunkett, his intellectual successor at Du Pont, who discovered Teflon® when fluorine was accidentally combined with ethylene in a laboratory fire; Elisha Graves Otis, who invented the first elevator that stopped itself if the cable snapped; and let us not forget Henry Ford, who, contrary to common misconception, did not invent the automobile, but did in fact invent and patent the spark plug.

You may have recognized most of the names just mentioned. But can you identify Frank Whittle and Hans von Ohain? They invented the jet engine and taught the world to fly without propellers. Marshall Goodrich? He proved that fluoride would reduce by half the number of cavities in your children's teeth. Wilson Greatbatch? He developed the first practical pacemaker to keep your Uncle Joe's heart beating regularly. Raymond Kurzweil? He made it possible for blind people to hear by an electronic scanner that "reads" the printed page aloud. Or, on a more mundane level, consider Herb Allen. He did wine lovers a favor by inventing a device to remove the cork from the bottle without strain, pain, or crumbling.

What distinguishes these people from many other inventors is that they not only changed the way we live today but made major money for themselves. And over and above the fact that such discoveries made possible new or improved, and often cheaper, products, inventions have historically been the driving force of the American economy.

Today, however, turning an innovation into a monetary miracle is no longer simply a matter of hard work and a modicum of luck. The moment of invention, that "eureka moment," is only the trigger for the business side of invention: the testing, the financing, the production scale-up, the marketing, the patenting, all the actions necessary to turn the discovery into a commercial success while simultaneously protecting the invention owner's position—in sum, the actions that turn creativity into cash.

Why many inventors fail to convert their original thinking into large bank accounts may relate to the design of the brain. People of intense creativity are operating at full power with the brain's right hemisphere, the side that is thought to generate inspirational ideas. Then, the practical issues arise. The crass confrontations of the business side demand attention from the less-exercised left hemisphere; and the result may be schizophrenia or, at least, mixed emotions.

Of course, this is not true of all inventors. There are geniuses who perform magnificently from either hemisphere. But millions from the mind are mostly made by inventors who know how to reach out and clutch someone to complement their own ingenuity with inventive ways of managing the business issues.

Because the words that people use in the world of invention are sometimes a bit confusing, here are some definitions—not according to Webster but as I intend them to be understood in this book.

Glossary

discovery The uncovering or observation of something previously unknown that may, as with chemical combinations, lead to an invention.

gadget, gilhickie, widget, device, contraption, novelty An item with some utility or titillation value, sometimes original, sometimes not, but seldom the foundation of a fortune.

idea The root of any innovation. Everybody has ideas, good, bad, and indifferent. Few people convert their ideas into actual inventions. Many people, upon seeing a new product enter the market, exclaim, "Oh, I had that idea years ago!" Such people usually make conversation but seldom make money.

innovation Something that hasn't been done before, or at least not recently. A *new* innovation is redundant; an *original* innovation is an oxymoron. This is a banner term that covers everything from an idea to an invention to a new product.

invention A tangible discovery or a creative insight, reduced to practice, that has utility, novelty, and feasibility. An invention can occur by accident or by forethought.

new A three-letter word that is often placed on labels because the designer had a small hole to fill. Although threadbare from use, it remains indispensable to marketing. What's really new? Is it a new flavor of an old product, a new-size package, a me-too product that's new to the maker's line, or a new "positioning" for the same product? In this book, I frequently use the reflexive noun "new-new" to connote a product that is functionally different from its predecessors or competitors.

new product In the lexicon of *Millions From the Mind,* any item that can promise the user new benefits based upon tangible inventions or discoveries.

Who Says It Can't Be Done?

Jack Rabinow, inventor of one of the first automatic mail sorting machines, the self-correction electric clock, and dozens of other elegant ideas, once told me, "Inventing is a pleasure; making money from inventions can be a pain." The reason for this is intrinsic to genuine innovations. When something hasn't been done before, when a product is truly new, there are unanswered questions: Can it be manufactured in volume? At what cost? What's a realistic sales forecast? How much money must be put at risk before these open questions are answered and profits start flowing in?

Yet enterprising people, both inside and outside major corporations, are still becoming rich, and sometimes famous, by exercising their inventive hemisphere and, one way or another, making all the right moves on the business side of invention.

Case in point: Back in 1984, an independent entrepreneur, Dick Berger, created the concept of a mouthwash that would be especially formulated to help remove plaque from the teeth. But who would ever want to go into the mouthwash business and compete against Scope and Listerine?

When he sought to license the product to a major marketer, company after company turned him down. But Dick made the moves that brought the product to life. To begin with, he came up with a great category-defining name: PLAX.

It is almost impossible to overestimate the value of near-perfect, preemptive trademarks like PLAX. They are hard to find and even harder to register because they are often simply descriptive. Yet the "right" trademark is so potent that it justifies the time and cost of an intensive search. A mark that is relatively easy to remember, visible when put on the product, suggestive of what the product does or, at least, what it is—in other words, a mark that helps sell the product even before millions are spent to make the name famous—such a mark makes each advertising dollar work harder, makes the competition's job difficult, and makes the company more valuable when you sell it.

With the antiplaque chemistry in place, Dick Berger paid fanatical attention to every aspect of the final product. Because people were expected to brush their teeth after using PLAX, he worked and reworked the formula to make the flavor compatible with every major brand of toothpaste. He revised the label again and again to make it a self-selling "billboard." Finally, in partnership with Norman Garfinkle, who provided the financing and the business structure, and thanks to the talents of Truman Susman and Joe Morano, ex-Bristol-Myers dynamos who did minor marketing miracles, a national distribution base was established for the product. No profits to that point, mind you, just losses.

Then, in 1987, Pfizer, Inc., bought 20 percent of Oral Research, Inc., for $20 million, providing critically needed working capital. Sales went from about $25 million a year to $100 million, thanks partly to Pfizer's own Leeming-Pacquin division's pushing PLAX, and in 1988 Pfizer bought the rest of Oral Research for approximately $200 million more.

Another entrepreneur, Charles Muench, was asked by two engineers to back the development of a color printer for computers. Back in 1982, Muench thought the market wasn't ready for that product. So he turned the basic invention into a copier that would make duplicates in four colors *and/or* black and white. Starting with a $200,000 investment, he raised the money to bring the Colorocs Universal Copier to fruition. Among Muench's right moves, he debugged his copier with surprising help from Xerox, enlisted Sharp in Japan to be his manufacturing arm, and captured enough Savin stock to turn that company into his U.S. sales arm. Ultimately, he overreached himself on the financial front and was forced out of his company, but not without a substantial increase in his personal wealth. That story is detailed in Chapter 2.

Inside major corporations, too, heroes are made, not born. At 3M Company, Spenser Silver discovered a strange adhesive that was sort of permanently tacky but never really bonded with anything. He spent five years wandering the halls of 3M's headquarters in St. Paul trying to find a commercial use for it. Then he linked up with Arthur Fry, who, one Sunday as he sang in his church choir and watched the little paper markers he had placed in his choir book flutter to the floor, thought at once about that only-slightly-sticky glue.

But even when Post-it™ Notes became a reality, it took a stroke of uncommon common sense from Geoffrey Nicholson, leader of the 3M venture group for Post-it, to turn the failing test markets into a $100-million business by remembering that the most effective tool for a genuinely new product is sampling. Silver, Fry, and Nicholson all made the right moves—proving that invention can survive and succeed against adversity in large corporations when there is the right corporate culture.

What Do the Winners Do Right?

The stories of PLAX, Colorocs, and Post-it Notes all have something in common: They were brilliant ideas that would have come to nothing without postinvention attention to the business side.

The inventors of those products, and the other winners whose stories appear in this book, had the capacity for taking infinite pains, which is one definition of genius. They fine-tuned the product, helped solve production problems, secured early- and second-stage financial underpinnings, and sought protection via patents, trademarks, secret processes, and other product safeguards. Some of them invented unconventional approaches to marketing or licensing. And all of them found allies for those areas in which they could not help themselves.

What separates the people who succeed with their innovations from those who—often with equally meritorious products or technologies—fall by the wayside?

While there are no universal road maps, if you consider the experiences of the fifty-one inventors chronicled in this book, coupled with my commentary and advice, you may well find your way through the rough patches and straight to the pot of gold.

There are two qualities, however, that you cannot acquire from this or any other book: judgment and belief.

You can sharpen your judgment a bit by studying actual experiences, your own and others'; but even though history repeats itself, it does so with variations, and these often prove critical.

Belief is different from judgment. Belief comes from within. Belief tells you that no matter what those dummies out there say, your innovation is wonderful and someday they'll all eat crow.

It's what made Spenser Silver hang in there for five years with his only-slightly-sticky glue until a commercial product was found. It's what made James (now Sir James) Black keep on searching for an ulcer cure until, long after SmithKline management had officially killed his project, Black discovered cimetidine, which became the billion-dollar drug Tagamet®.

Of course, there are those who pursue an innovation long after all signs of life have gone out of it and the rest of the world is crying "Enough!" For that reason, not all the "success stories" recounted here have a perfect happy ending.

Which brings us back to judgment.

Predicting Innovation Success: Tripp's Ten Tips

For evaluating an invention, balancing belief with judgment, I have developed certain parameters that help define success potential. You can apply these rules at any stage in the development of your technology or invention. I call this evaluation method the Success Indication Rating (SIR) scale, or, less stuffily, Tripp's Ten Tips.

You can rate an innovation, maintaining objectivity as best you can, by scoring 1 to 5 points for each criterion:

points

5 "Yes, absolutely, top-drawer, no problem!"
4 "Very likely, excellent probability."
3 "Realistic chance, some tangible reason to believe it."
2 "Maybe, at least it's logical."
1 "No way, José," or "I just don't know."

1. *There is a preexisting, well-recognized need for your innovation.* If people already know that they have a need for your invention, then getting a new product licensed to a company or bringing it to market is a lot easier than when you must create awareness of the problem before you can sell people your solution.

There are exceptions. Certain of the greatest inventions, especially those that create a whole new field or category, strike out on this score. Before the invention of radio, for example, most people were quite content with newspapers; before the invention of television, most people felt no urgent need to have pictures with their radio broadcasts.

But most of the time, it's easier to sell a Motrin® that helps relieve the daily pain of arthritis than an Ultralife® battery that promises to last ten years (if you live that long).

2. *The innovation is self-demonstrating, or at least self-evident.* The more you have to explain the innovation, the tougher the sell will be. Conversely, the easiest sell is when you can put the product into action, not say a word, and have everybody applaud.

As a kid, I sold can openers door to door. At that time, housewives were still using those lethal can cutters with a sharp point that you drove into the tin and then wiggled around the rim, producing a mean, jagged edge on which you could readily slice your finger.

I went around selling a neat little gadget that you snapped on to the side of the can. As you turned the winged handle, the cutter forced itself down into the can top and a knurled wheel moved the opener around the can rim, neatly and evenly separating the entire can top from the can body. I made so many sales that I returned my allowance to my parents for three months.

Look for the product with the self-evident advantage. Such products require less marketing effort, fewer advertising dollars. What's more, they continue to resell themselves to initial buyers, who sometimes become self-appointed missionaries for the product.

3. *There is an easy-to-discern difference that "stands in" for the major advantage.* When the product benefit is not immediately evident, it can be differentiated by "secondary characteristics." For example, Efferdent® Denture Cleanser was considered by its maker,

Warner-Lambert Co., to be an excellent product. But how many consumers can judge what's going on in a glass when their choppers are in there soaking? Warner-Lambert licensed an invention from Herbert Green that made the solution in the glass turn from blue to clear after about five minutes. They told buyers, "When the water turns clear, you know your dentures are clean."

Nor is the reinforcement necessarily confined to consumer products. Marquardt "super strong" fan belts come with a color stripe down the middle. Carpeting comes with a "Wear Tested" label. And the patent number or even "Patent Pending" label on a product generates credibility for the functional claims.

4. *No major capital expenditure is needed during the initial test periods.* One reason so few people go into the manufacture of such products as automobiles or toilet paper is that it's hard to assess the market without first making huge fixed investments.

As with most rules, it's the exception that illuminates it. Ralph Sarich, an Australian, has invented a new kind of automobile engine. He has licensed it to Ford, General Motors, and other automakers and made himself one of the four wealthiest men Down Under—before mass production has even begun. How Sarich convinced the biggest companies despite the Wankel engine disaster is told in Chapter 12.

Occasionally, you can substitute ideas for money to get past the initial test period. To make samples of plastic products, you can use relatively inexpensive aluminum or even rubber molds instead of steel molds. Orders can sometimes be taken from handmade models to generate working capital. Videotapes can substitute for the actual product during early buyer testing.

There's always a way to overcome what's missing. 3M's Art Fry, when faced with outlandish estimates of the cost of machinery to produce Post-it Notes, built a successful prototype machine in his own basement. (When it was finished—a classic inventor story—he couldn't get it through the basement door, and they had to break through a wall!)

5. *There is more than one way to cash in on your innovation.* Trust me on this. At the entrance to the new product maze, it's always possible to visualize the exit. Assuming that all goes well with the development work, the cost of goods, and so forth, how are you going to turn this creativity into cash? A good answer always envisions more than one solution. Let me reiterate: Even at the outset, a good "cash in" strategy always includes more than one solution.

If it's a technology you expect to license, you should be able to identify three or more companies that you believe will lust after this innovation. If it's a technology that makes possible several different products, preferably in different product categories, that's a big plus.

I once worked with a very bright inventor named Walton Smith on a technology that produced freeze-dried antacids and laxatives. It had such broad appeal that we licensed it to six different companies; and, even though the product was never a big winner in the marketplace, as licensors we made good money from the technology.

6. *There are layers of proprietary rights, so that you can go slowly in revealing them.* It isn't enough to file a patent application and then relax. If one patent is good, two are usually better, especially if one is still pending. If you can add the perfect trademark to your patent, the combination is like "one and one makes three" in terms of both protection *and* monetary value. And there are other techniques for setting up obstacles to deter either a licensee or a competitor from eating your proprietary lunch. Read Chapter 5, "Building Other Proprietary Picket Fences."

The other side of self-protection is how much you decide to reveal and when. Salome had the right idea with her Dance of the Seven Veils. Knowing where to draw the line is, of course, an art; but the basic principle is to keep your inmost secrets under wraps until the deal is signed and the check clears the bank. I once did an option-and-license with one of the world's largest pharmaceutical companies under which we furnished finished samples of *their* drug in *our* delivery system for them to test for six months—all without revealing to them how we put the system together. (Of course, their tests were *in vitro,* not *in vivo.*)

7. *Performance tests and customer acceptance tests show the potential of your product.* Whether you are starting a new company to launch an invention or aiming to license it, tests by third parties are relatively cheap compared to the value they add to the property. How much claim support you will need depends on whom you must convince—financial people, salespeople, consumers, or a potential licensee.

Shelf-life tests. Continuous use tests. Customer price-perception tests. What needs to be tested depends on many factors, as the chapters on marketing in Part III point out. But what you really want to know is, does the innovation's performance live up to your claims? Do people yawn or nod pleasantly, or do they challenge you? Or do they stand up and cheer? The most important test, of course, is when people pay real money to buy your product.

Even when you think the virtues are obvious, remember that, in addition to professional naysayers, most people—R&D folks, marketing mavens, business managers, legal eagles, and potential investors—are afraid to stick their necks out. Some people in corporations are more interested in how they *look* doing their job than in the results they achieve. Nor are they entirely to be blamed, because most managements punish failure far more than they reward success.

So test results that convince the doubters are important for innovators within the corporate world as well as for inventors/entrepreneurs.

Even with evidence of acceptance in hand, you still may not convince everyone that the product will fly. But it vastly improves your chances of finding one person who will become a champion for your cause; and sometimes that's all it takes.

8. *The product is not just new but newsworthy.* In today's marketplace, the word *new* is not new but old. If news is inherent in the product rather than in the word *new* in the press release, the innovation's chances of success increase geometrically.

As my friend Robert Siegel, former IBM'er and a master of the public relations craft, keeps reminding me, "Don't give me the whole history of the invention—just tell me, what's the headline?"

The best example of headline impact that I've ever seen was a United Technologies ad captioned "Keep It Simple" (see Figure 1). Yet oversimplifying is dangerous, too. Always remember what Albert Einstein said: "Things should be kept as simple as possible—but no simpler."

The right headline has two values: If you can tell the story in a few words, people will probably understand you; and if the news has any real impact, the product will launch faster under the "halo effect" of news than it would through advertising alone.

A good example is James Murdock's *Endless Pool,* an eight- by twelve-foot swimming pool in which the water moves continuously from one end to the other while the user swims in place. The *New York Times* caption read, "Watery Treadmill: Swimming Laps in a Machine." One of the tabloids put it, "Man Swims a Mile Without Moving an Inch!" Because it was newsworthy and made good headlines, in just one year Murdock's new pool was featured in fifteen major publications, including a story and full-color photo of me in *Fortune,* with the legend "Tripp, going nowhere." If Murdock had had the money to buy the same amount of advertising space, it wouldn't have had the same effect.

9. *You know your "time to the finish line" and are aware that the tortoise and the hare story is only a fairy tale.* In today's marketing and management climate, the fast track is not for yuppies only. Product response time has been shortened by computer engineering and computerized distribution. So innovation breeds competition, fast.

Figure 1. Ad from United Technologies' "Think" series.

Keep It Simple

Strike three.
Get your hand off my knee.
You're overdrawn.
Your horse won.
Yes.
No.
You have the account.
Walk.
Don't walk.
Mother's dead.
Basic events
require simple language.
Idiosyncratically euphuistic
eccentricities are the
promulgators of
triturable obfuscation.
What did you do last night?
Enter into a meaningful
romantic involvement
or
fall in love?
What did you have for
breakfast this morning?
The upper part of a hog's
hind leg with two oval
bodies encased in a shell
laid by a female bird
or
ham and eggs?
David Belasco, the great
American theatrical producer,
once said, "If you can't
write your idea on the
back of my calling
card,
you don't have a clear idea."

Source: © United Technologies Corporation 1979.

For example, the Cuisinart®, despite patents and a category-creating brand name, had the market to itself for only a few years.

So look closely and imaginatively at the path your innovation has to travel to market. There are four key elements that determine how fast you have to move: the size of the opportunity, the probable reaction time of the competition, the depth of proprietary protection, and the time required to get your product to full-scale marketing and a positive cash flow. Rate your innovation accordingly.

10. *The time is ripe for your product.* In the 1970s, no one even considered introducing a new product that claimed to help reduce cholesterol. Mention "oats" and people thought of horses. In the 1980s, with most consumers able to pronounce *cholesterol* and some able to understand its effects, oat bran products flooded the market.

The impetus for all this came from a pronouncement by the National Institutes of Health lowering the acceptable levels of cholesterol by some 20 percent. It was a government-sponsored trend that conditioned millions of people not only to change their eating habits but to be receptive to products ranging from Omega Oil to jogging machines.

Spotting a trend, a social habit, a medical opinion, an economic fact—anything significant that will lift your innovation above and beyond the humdrum of daily business—often spells the difference between new product success and failure.

I've applied this ten-point SIR scale to many products, and it's an effective way to assess an invention's potential early on. If your innovation has a total score of between 41 and 50, it's well worth pursuing. If it scores between 31 and 40, look hard at the weak points to find better answers and ways of changing the product. If it scores 30 or less, send the idea, the technology, the whole thing back to the drawing board—or to the circular file.

How Much Is This New Product Worth ...and to Whom?

In an era when star baseball, football, and basketball players are paid $1 million or more a year, when TV news anchors command $2 million, when financial wizards make millions just by transferring a company from one owner to another, it's high time for people who create new products, new businesses, new jobs, and new taxable profits to have a better shot at making millions from their minds.

But inventions do not translate into big money automatically. The business side of invention requires many kinds of skills and knowledge. It is a science, but it is also an art. So the biggest opportunities are for the entrepreneurs-at-heart, the imaginative risk takers who want to go for the gold and are willing to hang tough every step of the way.

Inventors—and their creative allies, lawyers, financiers, managers, marketers, and spouses—read on!

Chapter 2

Dealing With Money

Money is the engine of invention.

Money will not necessarily solve all the problems involved in turning an innovation into a profit-making product, but lack of capital will surely slow you down and quite possibly kill your chances of success. As one of my millionaire inventor friends used to say, "Money isn't everything, but it's way the hell ahead of whatever's in second place!"

The core of the problem in raising money to commercialize inventions is that you are asking people to bet upon the future, the unknown, the unproven, something that is truly new. Almost all the winners you will meet in this book found unique ways of raising rounds of money to bring their invention to the point of prodigious payout.

Of course, that's equally true whether you're raising venture capital for a start-up company or fighting for a budget allocation inside a major corporation. In fact, the most successful new product people I've met within big companies follow the principles of independent entrepreneurs—while carefully maneuvering along the constraints of organizational life.

Experience indicates that inventors are characteristically overoptimistic about how much money will be required to move from building the prototype to entering the market. In general, when it comes to imagination capital or venture capital, forget the dictum of famed architect Mies Van der Rohe, "Less is more," and remember the advice of Miss Piggy, "More is more."

Before the creative souls mentioned in this book made millions of dollars from their discoveries, each one of them found a way to raise the money needed to nurture the embryo along the long road to commercial success. Some raised millions, some struggled with a pittance—but no one got by on the sheer brilliance of the invention.

How much money will it take to go the distance? How much is needed each step along the way. What's a realistic reserve for grief?

These are tough questions, but you cannot raise substantial capital without having answered them. When you don't have hard answers yourself, reach out to people who have experience in each facet of the development and marketing sequence. A business plan that shows not only the cash requirements but the sources of the estimates is a lot more impressive than anybody's round numbers.

The Five Bridges to Success

As Somerset Maugham once said, "There is not much you can do with the five senses without the sixth, money."

No matter how much money you seek to raise stage by stage, you have to cross the five bridges that lie between the original "great discovery" and the point where major money starts flowing in:

1. *The Completion Bridge.* You must raise the discovery from the idea state to a reasonably complete product whose feasibility and patentability and desirability can be clearly appreciated by people who control the purse strings but who may not be "skilled in the art."

2. *The Protection Bridge.* You must establish the basis for defending the product against competition, at least in its early years. This could mean a patent application or any of the other proprietary picket fences described in this book.

3. *The Testing Bridge.* You must provide evidence, other than your own invincible faith, that the end product will perform as claimed and will be accepted, hopefully even embraced, by those who ultimately will be selling it or buying it.

4. *The Production Bridge.* You must provide specific answers as to manufacturability and the manufacturing source, whether internal or external, as well as the product cost.

5. *The Marketing Bridge.* You must assess the best route to market for the product. This holds true whether you are starting up a company, expanding your present company, or licensing a third party.

Each of the five bridges exacts a toll: money to buy prototypes, money for product testing, for market information and research, and for gathering a team of skilled professionals in different fields. I've yet to meet an inventor or entrepreneur who could do it all. The most successful ones don't even try.

To foresee the cost of crossing all five bridges is one difficult job, but thinking about what lies ahead and making (then revising) your best estimates is essential to raising money. The trick in budgeting is to hit a balance between what you hope it will cost and what you fear it might cost.

Keep in mind that estimating how much money you may need at each stage will dictate where you go for financing and whether you aim at starting your own company or at licensing the innovation. The bottom line is this: Sophisticated investors know that you can't foresee everything, so you have to pin down as much as you can, because nobody, but nobody, signs blank checks.

What Makes Investors Say Yes?

People with money, their own or someone else's, quite naturally worry about throwing money away, losing an investment by being imprudent. However, certain approaches influence people to put up money for innovation development and sometimes make them believers or, occasionally, even champions.

To understand investor psychology, first ask yourself, why should it be hard to raise money for something new and wonderful, something that bears promise of riches for those who back the invention? Because things go wrong—and experienced investors know it. Everyone has heard of Murphy's Law, but one venture capitalist offered me this sobering thought: "I operate according to O'Brien's Law. O'Brien says Murphy was an optimist."

So the first order of business in raising capital for an invention or new product is to squeeze out the doubt. Before the investor and his minions ask you the deal-killer questions, ask them of yourself and find the best answers you can. If there's an issue that you know will arise eventually, don't be afraid to raise it at the investor meeting and answer it while you still have the floor. It is far better to do that than to have the issue raised by some yup-and-coming wise guy when you're no longer there to answer the objections and resell your position.

Understanding where people are coming from is a basic tenet of selling. In raising money for innovations, recognize that most financial people and many corporate executives are actually uncomfortable when confronted with a product that is genuinely new, especially if the invention creates a whole new product category. It's much easier to think about stealing a share of the market in an existing category. It seems more secure to people investing money. Because it's quantifiable, it feels comfortable.

Thomas Alva Edison, according to his biographers, understood that selling stood on a par with inventing. Put this point to work. If, for example, you are seeking to license a large corporation, present your invention not just in terms of its esoteric wonders but in concrete terms that hit the exact person you're addressing: For the VP Marketing, talk about increased profit margins; for the VP Finance, stress rapid return on investment; for the president of the company, discuss how this business opportunity will make sense to the board of directors and to securities analysts and in the quarterly report to shareholders.

In other words, practice the art of salesmanship.

And never be discouraged by negative reactions in a meeting. Please remember, genuinely new products have a long history of cool receptions. Wasn't it our nineteenth president, Rutherford B. Hayes, who, after participating in a trial telephone conversation, observed, "That's an amazing invention, but who would ever want to use one of them?" And who among us, back in 1980, when first told of the FAX machine, would have said, "Now that's something I just couldn't live without. Where can I buy one?" The market research chapter, Chapter 8, will give you further insights on how to consumer-test products that are ahead of people's normal experience.

What It Takes to Raise Capital

Are there certain consistent characteristics in the scenarios of inventors or invention developers who have succeeded in raising early stage capital? Yes, indeed. Here's a summary of the hallmarks to emulate:

• You have a special relationship with a money decision maker. He or she owes you one, directly or indirectly. A common acceptance speech of inventors is "Thanks, Uncle Ned." As my son-in-law, Arnold Berman, M.D., the inventor of a superb hip-joint replacement, put it to me when he was looking for capital for a new project, "Save no favors for later... call in your markers when it will do some good." (In large corporations, this early money often comes from unused funds for other projects, especially toward the end of the fiscal year.)

• Your invention or product is self-demonstrating or self-evident.

The decision maker can see it happen, play with it, taste it, smell it, experience it, whatever. When Dr. Barrett Green invented carbonless carbon paper, any fool could see that it transferred the writing to the sheet below but that it didn't smear on your fingers. When you squirted Norman Ishler's aerosol discovery into water, the water immediately became effervescent; and when Lloyd Osipow and Dorothy Marra's aerosol expelled a small applicator pad, the pad was impregnated with whatever medicinal or cosmetic ingredient you'd put in the can. Even with jaundiced financial types, such self-demonstrating innovations tend to open the mind a crack.

• Your personal reputation, or that of your associates in the project, usually based on accomplishment in the field, convinces the decision maker that you'll be able to do it again. Classic examples of betting on the inventor abound in the heady climate of computers and computer software, especially in California, where bold venture capitalists like Montgomery Securities spearheaded the amoebalike growth of small electronics firms. Although sophisticated investors know that one success does not guarantee a repeat performance, if you or your group does not have a prior success halo, recruit a person who does. Investors love to hear, "This project is being done by the man who. . . ."

• You have more than one genuinely interested source of investment. In raising capital, as in selling or licensing an invention, it seldom pays to negotiate exclusively with one prospect—unless, of course, that source wants an option and is willing to pay for it.

• You radiate credibility. Ralph Waldo Emerson said, "Nothing great was ever achieved without enthusiasm." Not hype, but enthusiasm, the kind of deep-rooted belief in the rightness of the thing that becomes contagious and converts nonbelievers.

• You do not suffer delays gladly. A clean "no" is better than a repeated "maybe." There is a short story called "The Idol" in which a man builds a fantastic mechanical Buddha-like figure. He expects that, with something approaching artificial intelligence, one day it will spew forth the wisdom of the world. Many years later, as he sits before it, old, bearded, and bent, the idol suddenly shakes and roars "Time was, time is, time never again will be," and then self-destructs.

For those who champion an invention inside the corporate womb, the time pressure is less or may appear to be so. It's actually more an illusion based on the need for collegiality and mutual congratulations than a dictum from the top. When R&D people think that their company's pockets are endlessly deep, they may be reaching into a pocket with a hole in it. Even the most promising ideas can get killed when the project runs too long and money too short.

Rushing to judgment should not, however, foreclose the right of an idea person to doggedly pursue a treasured discovery. Spenser Silver, for example, trotted around the corridors of the 3M Company for almost five years trying to get support for the glue that made Post-its possible. Then he hooked up with Art Fry, who had just the right use for the glue. The details of that extraordinary story are revealed in several chapters of this book.

Of course, it's even tougher for the small laboratory, the smallish company, the university professor, and the individual, who often put everything at risk on "an only child." Not only must they work on tight budgets, but they seldom get to the second innovation if the first one fails, whereas new product people in corporations, where budgets are bigger, often have the opportunity to work on a number of innovations. So the choice in the smaller lab is to work harder, substitute ideas for money, and aim for one big winner.

Still, in the final analysis, within or outside corporations, the major tenet for successful innovators of new-new products must be: Make it happen now—before somebody tells you it can't be done.

Patents run for just seventeen years. Competition, however, runs day and night, every day, every year. If you approach investors sequentially and wait for each of them to make up their minds sequentially, or if you fight for the last percentage point in the deal, or if you let improvement after improvement delay getting the product to market, you may wind up like the idle idol.

Dramatizing the Opportunity

If you expect to raise money from any of the sources discussed in this book, you must first consolidate all the logical arguments and evidence to distinguish your innovation and its profit potential from pseudonew products. But logic and lab tests can be dull and uninspiring. You need to create some drama and excitement around the invention, bringing the opportunity to life.

David Montague invited bankers to ride his eighteen-speed bicycle and then fold it up and put it in the trunk of their car.

Charles Muench asked a financial man for a picture of his child and ran it through his copying machine, making a perfect black-and-white copy. Then just by touching two buttons, he ran it through again and made it come out in full color.

In short, involve people with your innovation. Don't just talk about an invention; make them care what happens to it.

Sometimes you can do this by creating a picture of a world in which the product is already accepted. A friend of mine sought to launch a new dental product for children. The presentation to venture capitalists consisted of one testimonial from a dental school and a ten-minute videotape of child after child using, reacting to, and talking about the product. The scientific reports made the investors believe that the product worked, but the kids' enthusiasm made them believe it would sell.

Gold at Opposite Ends of the Rainbow

People who do make millions from the mind use widely varying techniques to obtain financing.

Here are two examples, each the antithesis of the other in terms of financial sophistication, yet each highly successful for its innovator. The first is an almost miraculous start-up company with a single product in a major category. Time and again, the company required big infusions of capital to keep moving across the five bridges. The second example is an inventor of "small" products who created a big income stream with almost no capital investment. Together, they bracket the extremes of financing innovation.

First, consider the case of Colorocs Corporation and Charles Muench (pronounced Mu'-nick, like the German city). In moving from an idea to a going business, Muench invented solutions to problems that are universal in new businesses, even getting a competitor to help him. He became an eye-opening example of the inventor as entrepreneur—until he went a bridge too far and brought the wrath of the banks on his head.

Muench started Colorocs in Norcross, Georgia, with a $200,000 advance from his wife—about enough money to buy a nice house, but an insignificant sum to develop, make, and market a combination black-and-white and full-color copying machine. Yet all this was accomplished within seven years and Colorocs became a public company with international distribution of its Universal copier.

The genesis of the idea came to Muench from three engineers who had been working for Datapoint in San Antonio, Texas. Their concept was to make a full-color printer for computers using the principle of xerography for each of the necessary four colors but transferring the image via a series of endless belts—somewhat like an offset press—so that, when the last belt "kissed" the paper, both belt and paper were running parallel and absolutely flat, making possible accurate color registration.

After some study, Muench told the engineers that he thought the computer market wasn't ready for a color printer—and marketing uncertainties would make it hard to raise money—but that the technology might make a great color copying machine.

They started across bridge number one, aiming to prove that the idea was feasible. But the $200,000 seed money was not enough to design and build a prototype from scratch. So they modified several existing black-and-white copiers to prove the principle of the invention, namely that they could lay one color of toner on top of another and produce full-color reproduction in register. When one of the prototypes worked reasonably well, they were over the first bridge and ready for serious financing presentations.

Muench asked the engineers, "How much time and money will it take to develop the final machine?" Their estimate was three years and $5 million. Later events proved this estimate to be only one-third right; it took five years and $15 million to reach the production-ready stage.

At the beginning, Muench was still running Intelligent Systems, Inc., a company he had founded to make and market computer terminals and peripherals. He had taken that company public several years earlier, and a coterie of friends and acquaintances had made substantial money on the stock before small technology issues went out of vogue. Once the initial prototype copier proved feasible, Muench went to this group and raised $1.5 million to begin serious engineering of the product.

That sounds easy enough, but the elements that made it work are the ones everyone needs to raise imagination capital from individual investors: personal credibility and a tangible, easy-to-understand prototype or other evidence of progress. That let Muench get across bridge number one, completing the invention.

A year and a half later, with the first three prototypes at hand, Muench went back to the same people for additional money, and by the end of the third year they had invested $3 million in Colorocs.

While the engineering was moving ahead, Colorocs began to cross bridge number two, filing an array of patent applications. Even though xerography itself was an old art and even though the Xerox Corporation and other copier companies had hundreds of patents in the field, the specifics of the belt transfer system and the special kinds of color inks required for the process provided fertile ground for Colorocs's patent applications and for secret technology as well.

But time went by and the copier still wasn't production-ready. As with the hydra-headed monster, when one problem was solved, two others sprouted in its place. Then, just as matters grew really crunchy, a remarkable episode took place.

Charles Muench recounted to me how Xerox came into the picture:

> We opened conversations with Xerox on an open-end basis, and Xerox even invested $300,000 in us on a loan basis with the idea of converting it into equity later.
>
> Xerox sent people down here who analyzed our copier. They had seven Ph.D.'s down here, one for each area of the machine. We went through the Xerox labs showing our prototype. After weeks, they gave us a hundred-page report. When I saw that report, I said to myself, "Why did I ever get involved with this project? There are so many problems to be solved!"

Then, in an unexpected way, Xerox turned out to be Colorocs's best ally. After spending months on the project, Xerox suddenly notified Muench that it was pulling out because its lawyers were worried about possible antitrust problems. But, as Muench told me,

> They were very nice. In the hundred-page report, they had spelled out all the problems for us. Then—while they wouldn't give us their technology—they told us, "Ask this supplier if he can solve this problem for you" or "That manufacturer can probably help you on that question." As a result, we solved most of our technical glitches. The consulting advice was worth a quarter of a million dollars to us, and we didn't pay back their $300,000 loan until we got our next round of financing.

So bridge number three was crossed, thanks in part to a gracious competitor.

Of course, that's a dangerous game. Going to a top company in your field and exposing your invention-in-progress involves risks, especially when a similar product could be incubating in its labs. Still, Muench had assessed the situation correctly. Colorocs's patent position was a deterrent to outright rip-off and Xerox's high ethical reputation was comforting.

With the technical problems largely solved, Colorocs still had to cross bridge number four: Who would manufacture the machine? There were no American manufacturers of copiers other than Xerox, and raising enough capital for production engineering and a brand-new plant was next to impossible.

Muench traveled to Japan to seek out copier manufacturers—without success. Then he met a man in the local Atlanta office of Mitsubishi International Corporation and found an enthusiastic champion. Muench discovered what many others have learned, that the best way to deal with the Japanese is through a Japanese partner or an associate.

Mitsubishi helped Colorocs establish a master contract with the Sharp Corporation under which Sharp did all the production engineering *and* paid Colorocs $1.5 million to "equalize" Colorocs's prior engineering investments in return for manufacturing rights, exclusive distribution rights in Japan, and co-distribution in the United States, with Colorocs retaining rights for the rest of the world. The complicated deal evolved in phases, taking an excruciatingly long time, as most deals with the Japanese do; but it would never have happened at all without Mitsubishi, which not only ran interference in the negotiations but also bought $500,000 of Colorocs stock to keep the ball rolling while Muench sought to take the company public.

No major underwriter was interested in an initial public offering (IPO) for Colorocs, but Muench endlessly pursued new financial contacts until once again he found "acres of diamonds" in his own backyard. A local Atlanta firm, Marshall & Co., put the deal together in April 1986. With a better sense of timing than even Muench could have foreseen, the Sharp agreement was signed just before the public offering. That, together with Mitsubishi's $500,000, gave immense credibility to the issue and it was sold out promptly, putting $5 million in Colorocs's treasury.

Naturally, in the first year's rush to reach the market, much of the $5 million was burned up. Then production delays were encountered, including one "happy" problem: Colorocs engineers discovered a way to make half-tone copies as lifelike as fine lithography. This caused major retooling, but Sharp and Colorocs agreed that it was a strategic necessity.

Fortunately for Colorocs's survival, the initial public offering had two warrants attached, the first of $3.5 million of stock due one year after the IPO and the second for some $4 million, with the last exercise date just about a year later. Both warrants were fully exercised by the holders. These infusions of capital kept the company going through the preproduction period. They also gave Muench time to make a fantastic deal to gain working control of the Savin Corporation.

While Sharp and Colorocs engineers were getting the bugs out of the initial production, Muench had been seeking a way to cross bridge number five: a U.S. marketing channel for the copier. With Xerox a nonstarter, the logical choice for an alliance was the Savin Corporation. Unfortunately, Savin was having its own financial problems and every time Muench opened a dialogue with a Savin officer, two months later he'd find that the executive he'd been talking to was no longer with the company.

A complete financial rundown on Savin revealed an interesting item to Muench: Savin was 70 percent controlled by the Canadian Development Corporation.

The twists and turns of this story are too astounding for any words but Muench's own:

> I have a condo up in Toronto that we go to in the summertime, so I asked my broker friend up there about the Canadian Development Corporation. I'd assumed they were a government agency, that's how naive I was.
>
> Well, I found out that CDC was privately owned and the top management had just changed and the new people wanted to dump their Savin position. I thought, hmmm, here's an opportunity to get distribution for our copier.
>
> Now CDC's brokerage firm was McCloud, Young and Weir—MYW—and I told my broker to talk to their investment banking people. Well, they came on so damn aggressive, I couldn't believe it. Here we were a company with no revenues. We had only $4–5 million in cash and they said, "We can do the deal for you."
>
> I never understood until after the deal was complete why MYW, the fourth- or fifth- largest investment firm in Canada, took such an interest in Colorocs down in Norcross, Georgia, and put Colorocs and Savin together.
>
> It seems they had been trying to do a deal to buy a company named Silvermine from CDC together with Savin as a package, and they'd done a lot of homework on the deal. Suddenly, CDC spun off Silvermine to somebody else and McCloud, Young and Weir had done all that work and was stuck without a fee. So when I came by and was willing to hire them as an investment banker, they jumped on it—because they already knew a lot about Savin and I guess they figured, "Well, at least we'll get a $60,000 fee from Colorocs whether we put the deal together or not."
>
> The irony of the Savin negotiation is this: When we first talked with the people at Savin, they told us we were number twelve on a list of thirteen people who were inter-

ested in buying them. And the first name on the list was a company named Gestetner.

Now Gestetner is a hundred-year old English company that was acquired by an Australian group just before the crash of October 1987. At that time, CDC had been asking $80 million for Savin. After the crash, the Australians (through Gestetner) offered CDC $30 million in cash. CDC said no, and the Australians walked away. And so did almost everybody else.

With a relatively clear field, Colorocs negotiated a deal with CDC for $10 million in cash and $20 million in common and preferred stock for CDC's 70 percent interest in Savin. But when the stock market slumped in October, Savin's stock dived from 1¼ down to ¼. Concurrently, Colorocs's stock went down from 8½ to 3. So Colorocs's negotiators went for another adjustment: Before they would close on the deal, Savin had to restructure to eliminate a load of debt from its books. When this was done, Savin's net worth went from a negative $70 million to a positive $66 million.

When the deal finally closed, Colorocs got working control of Savin for $8 million in cash, a million shares of common stock, and 225,000 shares of preferred—with a total value of $27 million. The agreement was complicated; it included an arrangement to sell back some of the Savin shares later on, but it nailed down Savin's organization as the U.S. distributor for Colorocs copiers.

Muench owed much of that coup to McCloud, Young and Weir, who carried the burden of the negotiation and investment structuring. However, he was not too bad a negotiator himself. He persuaded MYW to take its $3 million fee in Colorocs stock, pleading that to pay cash would hurt his working capital position.

It took just one day short of a year to complete the deal and cross bridge number five.

Was it by luck or coincidence that Muench found his way through the financial woods? Or was it serendipity, the polite word that Maurice Hiles, a man who has made his millions from sorbothane and other materials, keeps using to describe his own success?

Without being unduly romantic, I'd say it's the quality I've observed in all the most successful innovators: an absolutely fierce unwillingness to take no for an answer. It's recognizing that the road to raising money isn't neat or predictable, and a willingness to change and adapt as conditions dictate. The winners always hang in until the law of averages gets a chance to operate. Like a good halfback in football, they keep on their feet and slide along behind the offensive line until they find an opening to dash through.

That quality produced for Charles Muench not only a six-figure income but stock in Colorocs with a multimillion-dollar market value. However, in common with many other inventor/ promoters, Muench went one bridge too far. To drive shipments faster, he introduced a program for putting Colorocs copiers into retail outlets like print shops without any upfront payment. The shop paid Colorocs 75 cents every time it hit the button. The company set up eleven branches to sell and service this business and offered the same pay-per-copy deal through Savin.

You guessed it, the impact on cash flow was murderous. In January of 1991, the directors of Colorocs relieved Charles Muench of his duties and installed a new president, Alan Srochi, previously executive vice-president, together with a new chief financial officer. Deborah Cox, head of public relations, neatly summed up the story: "I don't think anybody would argue with the fact that Charles was the driving force behind this company. I think that he fits the classic entrepreneur profile. You know, he gets it to the operating stage and then has trouble operating."

A Very Independent Inventor

At the opposite end from Charles Muench on the spectrum of invention financing is Robert Bennett of Easton, Connecticut. Bob is an independent inventor, a former patent draftsman for whom the time between perceiving a problem and solving it is frequently only seconds. With such fertile inventiveness at his command, Bob creates small, primarily plastic products or packaging components in his well-equipped basement shop.

Bob Bennett has no intention of crossing any bridges at all if he can help it. He spends as little as possible of his own time and money, operating in only two modes: Either he sees a problem, invents the solution, makes a prototype, and sells the invention to a company, or a company comes to Bob for solutions to specific plastic packaging or parts problems.

Bob's financial concept comes straight out of "take the money and run": Invest just enough time and material to create a working prototype, then make a deal with someone who will pay you as much money as possible up front plus a fee for completing or perfecting the invention. His pricing system is beautifully simple: Factoring in the buyer's apparent ability to pay and the probable time line to commercialization for the invention, ask for as much as you think the traffic will bear. Then take the first firm offer you get. And don't ever expect to get rich on future royalties; they're just gravy if and when they come.

When a company asks him to create to order, Bob takes a monthly fee while he's working on the development, followed by a substantial kicker payment if the company uses what he invents. Don't conclude from this that Bob is avoiding the business side of invention. He's just short-cutting the process. He knows the value of everything, including his own creativity.

For instance, when Bob created a one-piece, injection-molded dispensing cap for shampoo bottles and such, he told me that he priced it out right down to the last mill, took it to Revlon and sold it for $100,000 because he could demonstrate that they'd recover that sum in cap cost savings in about one year. Bennett's investment in time and materials was under $1,000.

You'd imagine that such a fertile mind would have produced at least one product so attractive that he would be tempted to start his own business to make and sell it. When I asked him about this, he protested, "No way. Too many headaches. Unless you count this fishing lure here that I sell by mail order and through a few local stores. Of course, this lure attracts bluefish from about a 100-mile radius—so I got quite a demand just from people hearing about it."

Whatever the merits of his fishing lure, Bob Bennett has little desire to raise capital and start a business. He told me that, year in and year out, when his accountant tells him that his income has crossed the $250,000 benchmark, he just eases off and goes fishing.

Reality in Raising Money

Most successful inventors land somewhere between the Charles Muench style and the Bob Bennett style of financing. And most of the time, it's a step-by-step process.

People who back innovation development are like patrons who invest in sculpture. They like to see what they're getting for their money. But there's a paradox: The inventor's vision is always a step ahead of the investor's.

Michelangelo once said that the figure already existed within the block of granite; he could see it in his mind and all he had to do was release it. But when he delivered to the duke a granite block with a half-emerged figure, the duke saw red and threatened grave consequences unless Michelangelo promptly released the rest of the figure from the stone. The serious inventor, like Michelangelo, can see both realities. The investor cannot.

The first and the last reality in raising money to commercialize inventions is this: What people with money want, with rare exceptions, is to make more money, with the least risk possible. If they were out to save mankind, they'd give money to the United Way, the Red Cross, or their favorite charity. If they had a yen for a seat on a board, they'd pick a going company and buy 5 or 10 percent of the stock. If they wanted safety, they'd buy U.S. Treasury bonds. But you'll seldom find "investing in inventions" on anybody's financial wish list.

So it's always necessary to build a bridge between the vision of the inventor/entrepreneur and those who are sitting on the cash. And this bridge must translate technology into cash flow.

Plainly, the need for imagination diminishes geometrically as the invention advances toward a running business. Thus, the potential investor who says no to the paper statement of the idea may say yes to a crude prototype. If the crude prototype still brings a no, a polished prototype or a short run of manufactured product may elicit a yes. Patent filings, claim allowances, or issued patents add value to the invention. Independent tests that show your product's superiority tilt the table your way. Add to this tests that show consumer or trade preference and investor interest climbs. If you can actually produce the product and put it into a sales test, get real orders, and make your initial deliveries, it becomes easier to attract not only investors but also those who would acquire the new product outright or acquire your company. Incidentally, a parallel progression exists for new product people fighting for budget allocations within a corporation: The closer the project is to market readiness, the easier it becomes to get money.

To package an invention as a money-making opportunity, rather than as a deep well for endless R&D and market-entry investments, you need a special kind of business plan.

Such a plan must not only cover all the bases of a business plan for an ongoing company, it must also spell out in believable detail how you expect to cross each of the five bridges and what it will cost.

Two very useful books on the basics for business plans are *The Arthur Young Business Plan Guide* by Siegel, Shultz, Ford, and Carney and *How to Create a Really Successful Business Plan* by David E. Gumpert. However, if you want an education in both how to write a business plan and what to *do* so that your start-up business succeeds, I highly recommend *Business Plans That Win $$$* by Stanley R. Rich and David E. Gumpert. Based on the experiences of companies that appear before the MIT Enterprise Forum, the advice is bedrock-practical. The man who runs the NEPA venture capital fund, Fred Beste, told me he insists that anyone submitting a proposal to NEPA read that book first and tag the recommended bases so Fred won't be wasting his time.

Even with such guidance, however, the real task is to make the future believable, and that is a function of how much specific support information you can gather to assure investors that you will, indeed, cross all five bridges. People who have money to allocate or invest have seen dozens or hundreds of business plans. They know that anyone can put numbers together on a spreadsheet. They want to know why they should believe yours. Finding ways to build credibility is the number one job of the business plan writer.

Walter Wriston, former chairman of Citicorp, tongue slightly in cheek, has pointed out that some people are so intent on safety that they think "risk is a four-letter word."*

How to Create Converts

At every stage of the development chain, turning an invention into a money machine is as much a matter of salesmanship as it is of engineering and patenting and testing and production. You need "converts," people who will keep their faith in your innovation when the problems start to emerge, people such as:

- The market research person who looks for a core group that loved the product inside a research report showing rejection by the majority.
- The patent lawyer who, with the prospect of a final rejection, hies himself to the patent office to shoot it out face to face with the patent examiner.
- The new product manager who bets the last $10,000 in his slush fund to add a last-minute improvement to your product.

Why do people do such things? Because they believe you have a winner. You have involved them, shared details, reported progress, injected a sense of excitement, and acted quickly to correct problems.

Even in the early developmental stages, you must picture an invention's marketing potential, differentiate it from the "me-too,

*Walter B. Wriston, *Risk and Other Four-Letter Words* (New York: Harper & Row, 1986).

more-too" school of "new and improved" products, and demonstrate its acceptance by potential buyers.

One key to creating converts is to focus your efforts on eliminating doubts. Just as everybody loves a winner, nobody wants to be associated with a loser. Look for the negatives. Either fix them or run a test to show that they don't matter to the buyer.

Of course, testing costs money. If you can't afford the required testing, try persuading your potential licensee or a future supplier or investor to pay for the tests in return for an option, with the understanding that the test results belong to you if the option is not exercised. Then, even if the deal falls through, you have something tangible to support your sales efforts for the next negotiation.

Obviously, the closer the product is to market readiness, the easier it is to get funding, but that can be a Catch-22. In Parts III and IV, the chapters dealing with marketing insights, licensing, and start-ups provide many ways out of this trap.

Letting the Investor Participate

Every experienced salesperson will tell you that getting the prospect involved with your product is more effective than a straight sales pitch. For example, the late, great Sam Prussin, inventor of Right Guard antiperspirant and other products, came to me with a radically new dentifrice, a clear viscous liquid that came in a small squeeze bottle. Sam had some evidence that it not only removed plaque but helped slow its re-formation, but he needed money for major tests at an independent dental clinic.

How could we persuade people to put money into an invention that was so far from the well-accepted tube of toothpaste, especially with only preliminary evidence of its functional superiority? As I tried the product in Sam's laboratory, he said, "When you finish, rinse and then slide your tongue over your teeth and tell me what you feel." My teeth felt supersmooth, almost frictionless. Was this the microcoating that would help prevent plaque? Good scientist that he was, Sam wouldn't say that without tests. So I told him, "Whatever's going on in my mouth, I know instantly that it *feels* different from any dentifrice I've ever used, and it *feels* like my teeth are cleaner than when I use toothpaste. That's the basis for letting people convince themselves that they should invest in this product."

We did a road show for a limited partnership, making presentations to West Coast folks ranging from the big shots of Beverly Hills to kiwi farmers north of San Francisco. After a brief introduction, every man and woman at the meeting was given a nice, fresh toothbrush and invited to the washrooms where our representative applied drops of the liquid dentifrice to their brushes and let them scrub away. Within weeks, that "convince yourself" method had raised close to a million dollars of venture money.

Whatever the field of invention, always look for a way to let the investor have hands-on experience. If you have developed a new, fast test for measuring the oil content in soil to determine drilling potential right in the field, take the venture capitalist out where there's an existing well and let him run the test. Or imagine the demo you could run with Herb Allen's ScrewPull® cork remover. Before Allen's invention, you twisted the corkscrew down into the cork and you yanked the cork *and* corkscrew out, either by brute force or with the help of some lever device. ScrewPull operates by letting the cork climb up a thin, stationary Teflon® corkscrew, eliminating the tug-of-war entirely.

If your product does not have this kind of visible advantage, create some tangible evidence to dramatize its uniqueness, something a purse-string person can understand. Invent a test that dramatizes its functionality; or make a video of potential buyers reacting to the product; or get a glamorous full-color illustration of the future finished product; or do a series of interviews showing that real people are aware of the problem your invention solves; or hire a trade paper writer to do a story about your breakthrough and have it set in type to show what kind of news it will soon create.

No matter what kind of product you are selling, there is one kind of convert you can always find—if you look hard enough and reach out—and that is the expert convert. This is a person who not only endorses what you have developed but who has the credentials to give the endorsement wallop.

Early in his launching of Chef's Choice® knife sharpener, Dan Friel recognized the importance of the expert convert, so he showed the device to many expert cooks and food writers, among them Craig Claiborne, longtime restaurant and food reviewer for the *New York Times* and an accomplished chef in his own right. Claiborne loved the product. Friel secured his blessing and the right to use his endorsement with his financial sources and in trade and consumer advertising.

* * * * * *

Once again, although I have addressed most of this chapter to the independent inventor/entrepreneur, the principles also apply to anyone who is charged with moving new products through a corporate environment to the marketplace. The need for more money, usually more than you expected; the need to create drama and excitement around the innovation; the need for pressing the timetable, for reacting quickly to change; and, most especially, the need for creating converts in every department including the financial decision makers—all these are urgent challenges within a going organization.

* * * * * *

Be creative after the invention. Nothing sells itself.

Chapter 3

Finding the Money Sources

Very early capital is, as you might expect, the hardest money to find. Very often, inventors have to look in their own pockets for it, entailing sacrifice for themselves and their families. Chester Carlson, whose inventions made Xerox possible, lived for several years in a small apartment where the kitchen was also the prototype development room, but his early sacrifices would have come to naught had he not found the Haloid Corporation and later the Battelle Institute to put up the money and manpower required to develop a marketable product.

Keeping the invention alive, of course, is the name of the game. Almost all innovators—save for those in big corporations, who usually find themselves going back to the same committee or board for the next-stage cash—find that they must go to a series of sources as their opportunity grows toward maturity. The secret of survival is to seek each successive level of financing *before you need it.*

The story of Stephen Bernard, who founded Cape Cod Potato Chips, is prototypical of the stages an inventor/entrepreneur goes through. As he recaps it,

> In 1980, we found some old equipment, including a kettle, and opened an 800-square-foot retail store. That's how we started, actually, selling across the counter in Hyannis, Massachusetts.
>
> We started out with $30,000, and this came from savings. We thought $30,000 was enough but soon learned that it wasn't. About a year later, we mortgaged the house

> to raise additional money for expansion. A couple of years later, we had some investors come in. As demand grew, we borrowed $100,000 from a bank and by the fourth year we got an Industrial Revenue Bond through the bank from the state of Massachusetts that let us put up a $3.5 million potato chip plant in Hyannis.

More details of Bernard's story are recounted in Chapters 5 and 8, although his sales promotion method was unusual. His financing progression, however, was typical of that followed by many successful entrepreneurs.

* * * * * *

Whether it's Stephen Bernard raising something less than $4 million or Marc Newkirk, whose story follows shortly, raising over $150 million, the money sources fall into four categories:

The Four Sources of Invention Financing

1. *The Bettors.* These are the first investors. Often, they are betting more on the inventor than on the invention. Their investment is low, but if the project succeeds, they get the biggest return on their judgment.

2. *The True Believers.* These private investors come in when the invention is reasonably complete, but some critical piece of the puzzle is still missing: Either the invention itself is not quite finished, or it has not really been tested, or there is no evidence as to marketability. The dollars are higher, but the participation per dollar makes the prospective return look exciting.

3. *The Deep Pockets.* At this stage, the invention is ready to become a business or has actually started, or major licensees have been signed. Professional investors and venture groups see that the technology works, that the market is there, and that the cash flow is about to happen or has begun.

4. *The Genuine Investors.* The product is on the market, but more money is still needed for increasing production or promoting sales. Now the conservative gatekeepers of pension fund money or the investment bankers may become interested. Or it may be time for a public stock issue.

When you are talking with money sources, it helps to sort them mentally into one of these four categories and to relate them to the state of development of your innovation. It also helps to identify the psychological reasons why each of these groups may invest—the need to be "right," the desire to be "in," the fascination with your field of invention—all the human driving forces that move people to back your invention rather than to buy stock in General Electric.

Money Is Where You Find It

If you are working on an innovation inside a corporate cocoon, you have only one way of getting money behind your project: you rally support from your confreres, write glowing reports, and agitate with key people in management.

If you are aiming to develop an innovation into a private fortune, whether through licensing or by starting your own company, you need a wide array of financial sources on your hit list.

Relatives and Friends

After your own savings and taking a second mortgage on your house, your father, mother, sister, brother, cousins, uncles, and aunts are the most likely source of additional money during the first phase of invention financing. Even though relatives can't resign from their relationship with you, advise them plainly, and in writing if possible, that their money may be lost. As you read in Chapter 2, Charles Muench started Colorocs with $200,000 funded by his wife. Others have started great inventions with less.

Friends, on the other hand, can be and often are lost when they lose money, so try to avoid small investments from friends unless you have a big circle of them. On the other hand, if one of your friends can invest substantial money, and you have unshakable faith in your invention, take the money and make your friend a partner.

Angels

In invention parlance, an angel is a person of substance who really backs an invention, usually somewhere between the Bettor and the True Believer stages.

A prime example of what the right angel can do can be seen in the case of Gordon Gould's laser patent. For eighteen years, Gould had been pursuing his patent application through the maze that is the Patent Office. One day he walked into the office of Gene Lang, president of Refac Corporation. As Gene Lang told me, "After looking at the patent application and satisfying myself as to the integrity of the man, and thinking of the challenge of being able to have a proprietary position in the use of light as an energy source, I decided to take it on."

Refac financed the pursuit of that patent for eleven years, first to get the patent granted, then to defend it against interference suits, challenges to its validity. The files are enormously thick but every suit was won, the last one being adjudicated in early 1988. The royalties from his laser patent will make Gordon Gould a very rich man indeed, but the angel won't do badly either: In 1988 Refac's royalties from laser licenses were in excess of $1 million. There's more about Gould's legal war in Chapter 6.

Limited Partnerships

Commonly called an LP, the limited partnership is a widely used vehicle for raising early-stage capital from True Believers. If investors like the invention, they are likely to like the LP format, which usually provides for early repayment of investors' money when income begins to flow, substantial profit kickers, and reasonable protection for the investors from liability. While the exact terms of each LP differ, most lawyers with a corporate finance practice know the variations and have in their computers the legal babble necessary to go out and raise money under an LP—and you've got to use them and not just copy somebody else's LP contract, because the technical legal requirements vary from state to state and keep changing. Dozens of Silicon Valley deals and other technology developments have been done this way.

If you know the first two or three people who will invest in your LP, it's worth having the paperwork put together so that you can get out and peddle the plan wherever you can. Just remember that the limited partnership has to have a general partner and that this GP has to run it like a business—or the limited partners may get madder than hell and sue. Even though it's customary to create a corporation to be the GP, if you're the person running it, you can't operate it as though it were your own money without risking complaints or even litigation.

Notwithstanding some of the problems experienced by real estate and oil limited partnerships, the LP is still probably the most feasible vehicle for raising capital for start-up technology-based companies. It certainly appeals to people with some wealth who like to bet on their own judgment rather than on the latest hot recommendation from their stockbroker.

Private Placements

If you are operating a corporation to develop your product, additional capital infusions can be generated by the "private" sale of securities to individuals or other corporations. You're looking for investors who range from True Believers to Deep Pockets.

One of the more ingenious formats for raising capital from individual investors was devised by William Greene, the man who invented and marketed the Tennis Tutor. This is a tennis ball machine that operates on batteries. No cord, plug, or outlet is needed, so you can take it anywhere. And it's silent. No popping of the ball disturbs neighboring players.

Bill Greene, who has an M.B.A. in entrepreneurship from the University of California, invented the Tennis Tutor to teach himself to play better. While still working as an electrical engineer for Lockheed, he raised $50,000 from friends to create an engineering plan and build a finished working model.

Then, he set up a "sub-S" corporation—the kind where both profits and losses flow directly through the corporation to stockholders in proportion to their holdings. Bill tapped a few friends for initial stock purchases.

Then came Bill's brainstorm: "I went to a friend who runs the biggest tennis facility in our area. He knew which of his clients had money, which ones had private courts. So very discreetly, whenever a friend would call in for a game, he would mention the Tennis Tutor to them and, if they were interested, we invited them over to hit with it. That's all."

The people who came to hit quickly bought a package deal: They became both stockholders and Tennis Tutor owners. And so did some of the bystanders. Bill Greene raised $150,000 and had the nucleus of a going business. Sales soon reached seven figures.

Any time you sell stock to third parties at arm's length, there are legal requirements as to the number of investors and investor qualifications, not to mention the numerous WARNINGS IN CAPITAL LETTERS that must be included regarding the dangers of the investment. So qualified legal counsel is an absolute must.

Once you're past friends and family, you'll find that getting help in raising private investment money is critical. *Whom* you know is important to any money-raising effort, but it is indispensable when it comes to major money for an innovation.

At the opposite end of the size spectrum from Bill Greene and his corporation are Marc Newkirk and the Lanxide Corporation. As this technology for making new basic materials from ceramic and aluminum developed, the need for more personnel and facilities escalated astronomically.

Every week, new evidence developed of the remarkable ability of Lanxide materials to resist erosion, heat, and wear under the most extreme conditions. In addition, new proof came forth that dimensionally accurate products could be produced even as the ceramic-metal material was being formed. As each new kind of matter was created and a product employing it was produced, new patents were filed worldwide.

Marc Newkirk had started with a vision of the interaction of aluminum and ceramic under certain conditions. Through Robert Roth, a stockbroker in Gladwyne, Pennsylvania, he was introduced to Bentley Blum and Paul Hannesson, two private investors who bought into the vision with $3.5 million, enough to prove the basic concept.

Newkirk saw the need not only for major money but also for the depth of engineering skill that large companies build up over the years. The first breakthrough came in 1985 when Alcan, the Canadian-based aluminum company, entered a joint venture with Lanxide with an initial investment of $18 million to develop specific uses of the technology. This was followed by other Alcan/Lanxide ventures and a total investment of $90 million. In 1987, Du Pont joined with Lanxide in joint ventures that brought an additional $82 million of capital into the orbit.

But Lanxide needed money to perform under those joint ventures and to do further R&D. As the product outlook improved, the need for cash grew even faster.

Robert Roth had introduced Marc Newkirk to Gerald Kaminsky, a senior officer at Cowen & Co. of New York, and Kaminsky and other officers of Cowen bought into the company. Cowen and Alex. Brown & Sons of Baltimore, both medium-size investment banking firms with worldwide connections, undertook to raise $11 million for Lanxide from private investors. Brown struck out, but Cowen raised it all.

Newkirk and Gerry Kaminsky, an executive vice-president of Cowen who had visualized at an early stage the immense potential of "engineered materials," worked as a team making presentations to big private-money people in the United States, Europe, and Japan. This produced the $11 million in stock purchases during 1989. When I asked Kaminsky to explain to me how they had achieved credibility for this radical new technology with tough-minded, experienced investors, he explained, "It wasn't an easy sell because Lanxide was such a new idea. But even when people didn't understand the process, they could see the dramatic results in the sample products. And remember, Newkirk is a believable guy."

In 1990, this time teamed with Smith, Barney & Co., Cowen delivered to Lanxide an additional $30 million of stock purchases.

By mid-1991, the Lanxide Corporation had raised from private investors and from its joint ventures, in cash and contributed services, a total of some $300 million. The first $100 million went solely to developing the technology to the point where products could be made. With 1991 sales of about $4 million, the company expects to reach the break-even point in 1993.

Marc Newkirk told me that a friend had given him this definition of an entrepreneur: "A man with a poorly developed sense of fear." By that standard, Newkirk is one helluva entrepreneur.

Venture Capitalists

The words *venture capitalist* (VC) loosely cover anybody who puts money into the other fellow's business in return for a share of the action... and sometimes other considerations. Some people think the phrase is an oxymoron.

Venture capitalists tend to accept only one in a hundred—give or take a few—of the proposals that they look at. No doubt, that's why there are so many people ready to take a cut at them. I once introduced an Englishman to a U.S. venture capital firm and when I asked him how he'd made out, he replied with only four words, "Capital fellows, no venture."

In addition, many venture capitalists have been known to demand very large percentages of young companies in return for desperately needed investments, leading to the appellation "vulture capitalists." You have to appreciate the VC's position, however. Typically, they make money on only one investment out of five start-ups. With that kind of batting average, they *must* get a big percentage of the action from the winners. (On the other hand, do you want to get into a deal with a businessperson who makes only one correct decision out of every five?)

There is approximately $160 billion of venture capital available in the United States. You might think that this would make venture capital the prime source for generating new technology-based companies, but there's a fundamental problem.

"The problem is," explained Alan Toffler, managing partner of Ivanhoe Capital of La Jolla, California, "that venture capital people move with the investing trends. There's an in time for the computer start-ups, there's an in time for software, then biotech is hot, then only second-stage financing or bridge financing becomes the thing to do. The result is, after a while, too many VCs are chasing too few good deals."

From the viewpoint of the entrepreneur/inventor, as we enter the 1990s, most venture capitalists are turned off by start-ups. They're looking for second-stage opportunities, where you've built a going business and need money to grow to the next level. VCs also favor spin-offs from large corporations of divisions or brands, a base for a business to which innovative products can be added.

None of the above should deter you from cultivating venture capital contacts. Like other human beings, venture capitalists come in a wide assortment of sizes, shapes, fields of interest, and amounts of capital available. And they are in touch with all kinds of people who may help you if they can't. Sorting out which venture capitalist to approach is your first job, and not an easy one.

You won't find a list of venture capitalists in your local Yellow Pages. Probably the most complete list is *Pratt's Guide to Venture Capital Sources,* albeit Levy's *Inventing and Patenting Sourcebook* and some other inventor-oriented books also include lists of venture capitalist names. Such lists often specify the business fields in which each firm specializes; these areas of interest are not necessarily current, but they're a start.

As you might expect, it usually takes more than a cold phone call to generate investment money. What you need are points of contact and, most likely, guidance. One good way to initiate that is to get to know people who have ongoing relationships with the venture capital community.

This contact point could be your bank. Some of the large banks have their own venture funds with managers who, if they can't help you directly, can point you to the right people.

Also, like other people in business, venture capitalists belong to associations. It may be a venture capitalists association, the Licensing Executives Society, or simply part of the Chamber of Commerce. Getting to the meetings of these associations is a prime way of getting to know the individuals you'll need when you're ready to ask for investments.

Another valuable source of help can be found at university business schools. Within such schools, there is almost always a course on entrepreneurialism, whether under that name or not. At some schools, such as the University of Pennsylvania, there are even separate schools of entrepreneurialism. Get in touch with the professor who teaches the course on the subject or the head of the school.

Ask for advice, or hire him or a graduate student to help write your business plan.

Your Banker

Having mentioned banks as a possible referral point for finding investment capital, I must quickly add that (1) they are almost certainly not a source of direct investment money for any company, let alone invention-based start-ups, and (2) they are probably not a source of loans either. This is not a criticism of bankers. I have money in some banks, and I worry a lot about the loans they *do* make. I wouldn't want to trust my bankers' ability to judge invention-based ventures and to lend them my hard-earned deposits.

So where does an innovator go for loan capital? Let me share with you the story of what happened to Noel and Adele Zeller, the folks who developed the Itty Bitty Book Light, that charming little halogen clip-on with articulated arms for reading in bed without precipitating a divorce.

At the time of this invention, their company, Zelco, Inc., had already launched flashlight innovations and a fluorescent lantern that won a place in the Museum of Modern Art's good design collection. Sales were running about $1.8 million. Then they introduced the book light, and the store buyers went crazy about it.

As Noel Zeller described it to me,

> Our bank at the time was Citibank. We weren't prepared to go from a million-eight to ten or twelve million dollars overnight. I presented the model to our account manager at the bank. I told him, "Look, I have a relationship with you guys. I have this product, I don't know how big it's going to be, but it's going to be good. I need $400,000 to $500,000."
>
> So he said, "On the basis of your balance, we can give you, maybe, another $100,000, but not $400,000 or $500,000. Furthermore, I can't make this decision. It has to go upstairs."
>
> They had a senior vice-president named Pat Goldstein in charge of commercial banking in Westchester County. And the Citibank manager said the only time she could meet me was at the bank's private box at Yankee Stadium the following Wednesday night. I figured she was going to talk to me personally and quietly, but she asked everybody in the box what they thought of the idea. Then she said, "I think this is terrific," and, around the seventh inning, she said, "OK. You have the money."

Based upon the bank's commitment, Noel Zeller went off to the Orient to place orders for the plastic molds and raw materials. When he returned, he got a call to come in to the bank.

"They told me they had made a mistake," Zeller said plaintively. "Their officer hadn't done her homework, the business was too small, the loan would violate banking regulations."

In the meantime, orders were pouring in. At one account, the bookstore chain Barnes and Noble/B. Dalton, the president, Lenny Reggio, and a bright buyer, Jeanette Limoges, started promoting the Itty Bitty Book Light in advance of delivery. They took orders for 10,000 the first week. When Noel and Adele Zeller told them of their capital bind, Reggio gave them a check for $150,000, saying, "This is money in advance of purchase, because we believe you are going to do it."

But Zeller still wasn't off the hook. He borrowed money from his agents in Canada and from some friends in Hong Kong. He was still short the last $200,000. As Zeller remembered,

> I was telling the story to an accountant I know, and this guy said to me, "There are ways of getting the money but you have to know that these kinds of people who lend this money get very high interest rates and if you don't pay back———you better bear that in mind." So they brought this fellow down who was a lawyer, and he looked us over and said we'd have to give him four or five thousand as a finder's fee and that he would get someone to lend us the money.
>
> Next thing you know, a guy comes walking in. We called him "Bigfoot"! He lent us the money at 28 percent interest a month. And the inference was, if we didn't pay it back, he would break our kneecaps or something.
>
> Somehow we muddled through the whole deal. The item sold well. There was enough profit that we paid back all the loans and had a cash flow to keep us going. That's the key to it. You have to have enough margin to keep going!

The inclusion of Noel Zeller's cautionary tale is not to be considered an endorsement of loan sharks but rather as evidence supporting my earlier comments on the paucity of support for innovation from banks, the normal channel for business loans.

Your Stockbroker

Sometimes your personal customer's broker may be helpful in guiding you to private investors or in introducing you to the key people in the underwriting department or in the investment banking department of the brokerage firm. He or she might even invest in your company personally, but don't build your whole future around this hope.

Government Agencies

Grants and loans from federal, state, or city agencies are a realistic source of money for start-up companies. Just be aware that government actions don't happen quickly.

At the federal level, the money is targeted at improving the United States' technological competitiveness or at solving national problems such as the cost of energy or pollution.

The Small Business Administration, working through eleven federal agencies (National Institutes of Health, Department of Commerce, and so on), makes annual grants for the development of innovations. The program is called Small Business Innovation Research (SBIR). Phase I grants of up to $50,000 fund feasibility studies, and Phase II grants of up to $500,000 underwrite full product development.

When Armando Cuervo's baby son developed colic, he conceived a sound-and-vibration device that would mimic the motion of riding in a car and quiet the child down. At the time, Cuervo was director of international operations for Ashland Oil Co. and had no experience in medical devices. He knew, of course, that the product would require testing, and a friendly pediatrician sent him a grant application from the National Institutes of Health. Cuervo called the director of the NIH division that dealt with pediatric products, a man named Dr. Alexander, and explained what he was trying to do. Dr. Alexander invited him to an SBIR seminar. Cuervo explains what happened next:

> A week later, I was in an auditorium with three hundred Ph.D.'s. I was the only non-Ph.D. there. But I found out how SBIR grants work. I came home and rounded up a team from Ohio State and from local industry to help me put that application together. We sent it to NIH and it was rated "almost perfect" by a panel of scientists. We were awarded a Phase I grant of $50,000.

That money paid for the clinical trials and early product development work. Armed with positive results, Armando Cuervo applied for a Phase II grant and was awarded $500,000 for a two-year period—enough to get his business off the ground. The product was named SleepTight®, and you'll find the story of its unusual marketing history in Chapter 8.

States and cities also have grant and loan programs. The target is usually increasing employment, that is to say, increasing tax revenues. Such programs abound; seek them out through your banker, accountant, or lawyer.

One example is the Benjamin Franklin Partnership in Philadelphia, a joint effort of the State of Pennsylvania and local southeastern Pennsylvania people. The Partnership makes grants for early-stage work and loans for business development. A representative case is James Murdock, who was manufacturing a unique product, the Endless Pool®. This is a small swimming pool that moves the water continuously from one end to the other and then recirculates it. You control the speed of the water and swim in place. Murdock needed to change from custom manufacture to a product that could be delivered as a kit and assembled in a few hours. In addition, Murdock had adapted the propulsion unit of the Endless Pool so that it could be installed on any small existing pool, and he called this The Fast Lane. The Ben Franklin Partnership not only saw an innovative product but also added jobs in Murdock's small plant near Philadelphia and loaned him $100,000. At least as important as this money was the added credibility gained. Shortly thereafter, James Murdock raised many times the amount of the loan from private investors.

Investment Bankers

With their own money or with the money they can raise from their networks, investment bankers are the prime source of funding for major developments. Note the cases of Lanxide and Colorocs already discussed. Investment bankers will seldom help small or start-up operations, but in any case, the time to cultivate your investment banking connection is before you need the money.

Underwriters and IPOs

Often investment bankers and underwriters for initial public offerings are the same people, just plucking different strings on their financial fiddle. But among the brokerage houses that promote IPOs, you'll find a wide variety, from giant Merrill Lynch to aggressive start-up launchers like D. H. Blair through to Denver's famous and infamous "penny stock bucket shops." For the financial-world novice, it's impossible to sort out the wisest path to going public. The best move is to hire an *ex*-Wall Streeter with underwriting experience. Put him or her on your side and, if possible, tie his or her compensation to stock options, exercisable upon a successful public offering.

Two small warnings about initial public offerings. Most underwriters will make it a condition that they have the right to do your future stock offerings as well. It sounds fair, but watch the exact language so that, if they don't perform, you can escape. Furthermore, note that, even though the letter of commitment may say that the underwriter will use his or her "best efforts," many underwriters don't actually *sign* the letter until the night before the offering—while you're on the hook for the costs and public embarrassment if the underwriter backs out.

There is a way to become a publicly held company instantly. You can buy a "shell" corporation, one that is already publicly subscribed but is not an active, operating company. Such companies may even have cash in the bank. However, the chances are small that you will find such a shell with a clean SEC record *and* then buy it at a reasonable price while keeping clear control *and* then be able to sell *additional* public shares to raise the capital that you needed before you started this exercise.

Partners

There are partners in the form of Deep Pockets, who may or may not become active in your business. There are partners who may contribute more in the form of knowledge or connections than money. There are partners who are really strategic alliances such as the joint ventures set up by Lanxide with Alcan and Du Pont.

Whatever the nature of your partnership, it will be a very different experience from raising capital through any other route. As in marriage, it is a joining together not only of material interests but of people with diverse personalities, corporate cultures, and standards. For that reason, financing via a partnership should be looked at from every possible viewpoint before the papers are signed and sealed. So an "engagement" or "living together" period, if it can be arranged, usually helps both parties sweeten the deal before the knot is tied.

Finding the right partner is the central issue. It requires an open, imaginative mind and a wide outreach. No case that I know of illustrates this better than Biomagnetic Technologies, Inc., a San

Diego company that has developed magnetic source imaging equipment, called MSI. In simplified terms, MSI uses ultrasensitive, low-noise amplifiers to detect changes in the minute magnetic fields that surround nerve activity in the human body. These amplifiers are actually superconducting quantum interference devices (engineers call them *squids*). Glamorous, right? But also tremendously exciting in terms of a doctor's ability to read what is going on in the human brain and other parts of the sensory system.

Such developments take major money. Dr. William C. Black, senior vice-president for business development at Biomagnetic Technologies, Inc., told me how BTI found an unexpected foreign financial partner.

After raising $9 million of venture capital between 1984 and 1987 and an additional $5 million in 1988, BTI found itself with its focus on medical products and its first FDA-approved device, a neuromagnetomometer. And the need for a lot more money.

As Bill Black described it,

> We approached leading medical companies first to see if they wanted to partner with us or assist us in marketing this product. And we found very little interest on their part. As it turned out, we ultimately made a deal with Sumitomo of Japan, even though they were not in the medical device business at all.
>
> We'd had some bad experiences with Japanese businessmen in the past because they were only interested in getting their toe in the door and learning about the business and we would have maybe two years before they decided to manufacture it themselves and we would be left with the role of advisers.
>
> It happened that Seito Trading Co. in Japan learned about us because they had put money into a venture capital group that invested in BTI. Steve James, BTI's president, talked to one of their people in New York and became good friends, and *he* went to bat for us, networking among Japanese businessmen until he came up with a most unlikely candidate, Sumitomo, the giant heavy-industry complex.
>
> As fate would have it, Sumitomo fit with us on two unexpected counts: One was that their base business was slacking and facing competition from other Asian countries. And two was that they had established a new busi-

> ness division to look for opportunities to diversify and they had a fair amount of cash to accomplish this.

Sumitomo had, in fact, already obtained Japanese distribution rights to certain U.S.-made medical products. So the BTI people made their position clear from the outset. They explained that they were interested only in a long-term relationship. They were looking not just for a distribution deal but for a substantial minority investor with a sufficient stake to have a vested interest in BTI's success. They wanted not just a distributor in Japan but a partner who would commit human resources to developing the market for the products, dealing with government regulatory issues, and servicing the equipment after it was sold.

That tough-minded approach appealed to the Japanese. Then they discovered another commonality. The MSI equipment depends upon a half-million-dollar metal-shielded room. Most of that cost is in the special alloy of nickel and iron and trace metals that form the shield. BTI had been buying the material from a subsidiary of Siemens in Germany, and Siemens was one of BTI's principal competitors. Although the use of this metal was incidental to the basic MSI technology, Sumitomo immediately saw that it could provide this shielding from one of its metal production companies. Just consider what Sumitomo was seeing: a leading-edge medical instrument invention, a forward step in superconductivity, and a fortunate match in special metals.

Nevertheless, as Bill Black carefully pointed out, it took two and a half years to complete the deal. And it almost didn't happen at all.

Steve James, Bill Black, and the officers of BTI had not been idle. They had been preparing for a public stock offering. When the talks with Sumitomo reached an impasse, they announced that they were withdrawing from the talks because they were in the midst of a public offering. Stunned, the Sumitomo people came back and decided that things could be worked out after all. And they paid a premium on the price. They bought $9 million of the $13-million public issue and that gave them 13 percent of BTI.

Bill Black summed up the learning experience this way: "I was astonished how difficult it was to communicate initially. But then you learn to adapt. Being there in person and getting to know these people on a personal level and developing a lot of trust and not being afraid to verify that they know what you mean and also the other way around... that level of communication is essential."

That is valuable counsel for any partnership in the making,

whether it is with a foreign company or a domestic one or simply with a single new partner named Joe.

All the Friends You Can Get

In some situations, you can find help from a variety of less obvious sources. For example, if you need testing to validate or finalize your invention, universities and commercial testing laboratories are filled with talented people who may help you in return for a small percentage of future income. Potential suppliers of material or manufacturing for your product can be enlisted for trial runs or, occasionally, investments. If you are still seeking funds when you reach the first license negotiation, consider asking the licensee for a large sum up front in return for not only an option on your invention but also for a portion of the future royalty stream from other licensees of the technology, in this country or abroad.

Whether you are going to The Bettors, The True Believers, The Deep Pockets, or The Genuine Investors, making the best possible presentation is critical. Please don't expect other people to envision your invention as you do. A horse, for example, viewed from directly overhead, looks a lot like a violin. But one needs imagination to see this. So dramatize your invention, spell it out, make it come to life. Even if you're a conservative scientist, it's your job to stretch the imagination of the investors.

One final word of advice on funding: Go for more than you need, not only because things usually cost more than people plan on but because money raising takes a lot of your time away from other things you'd rather be doing. As Bill Black said to me in a parting shot, "The effort of raising money is exhausting, takes a lot of persistence, dedication, hard work, willingness to talk to a lot of people—and you've got to be a salesman."

The amazing part is that so many people do find the capital to go the distance, use patents and other forms of protection, and license their inventions or start their own companies. That's what the next two sections of this book are all about.

Part II

Protection: Guarding the Gold Mine

Chapter 4

Using Patents as a Business Weapon

When I came home after four years of military service, I asked my esteemed father-in-law, Jack Beresin—who ran a large company that sold popcorn and refreshments in theaters and stadiums—what he thought I should do for a living.

"Son," he said, his blue eyes twinkling at their charming best, "get yourself a patent or a privilege."

There was a pregnant pause as I waited for the rest of the deep wisdom. "That's it?" I asked.

"Think about it," he said, "It'll make sense to you by and by."

What he was saying, of course, was that all business is rough and competitive. Whatever field you get into, find some kind of protected position. Jack's own company sold popcorn in movie theaters and hot dogs in places like the Houston Astrodome, where, as we all know, if you want what they've got, you pay what they charge.

Certainly, if you expect to make money from an idea—an invention, a technology, a new product—getting "a patent or a privilege" has to be a high priority. Think of it as buying time to get your baby on its feet, to stumble, to soak up the start-up costs, to make mistakes until it grows into a marketing force that can fight its own fight against copycats or outright theft. And keep on making money by keeping competition at bay as long as possible.

Patents are one way to protect products and processes. But before getting into the ins and outs of patents, please note that there are many other kinds of protection. The more of these you can combine with a patent, the stronger your position will be.

- *Trademarks,* especially trademarks that are a natural for the product and trademarks that become synonymous with (but not generic for) a new category.

• *Copyrights,* an often-neglected form of protection that can add an extra picket to your fence.

• *Trade secrets,* the know-how that makes the difference between a passable product and a great one, which is sometimes hard to guard but invaluable in the sale or licensing of a product.

• *Supply control,* that is, retaining the manufacturing of the product or the sourcing of a key ingredient, especially when someone else is doing the marketing.

• *Third-party endorsement,* that is, using the influence of people who count with the buyer (all the way from Michael Jordan's endorsement of "the pump" athletic shoes to a certain drug your doctor prescribed because *The New England Journal of Medicine* reported on it.)

• *Government regulatory approval,* an often onerous problem, but also a shield when your product has approval from the FDA, EPA, or some other regulatory body.

• *Publicity and advertising,* tools you can use not only to preempt the competition but also to threaten them (see the story of Picturesque® stockings in Chapter 5).

• *The law of the land,* using a variety of legal measures relating to unfair competition, theft of intellectual property, common law rights, and so forth, which, given astute lawyers, may help protect you against thievery from within or without.

Each of the above product protectors is covered in some detail in the next chapter. But first I want to talk about what patents can and can't do for you and how to make the right moves in this sometimes arcane field.

Using Patents for All They're Worth

In the United States, as well as in most but not all industrialized countries, the patent system is your primary line of defense because it says plainly, "Keep off my turf!"

Some people, however, don't think much of patents. When I first met Bill Valentine, the man who invented a pharmaceutical tablet that turns to liquid when you chew it, he remarked, "Patents only give your competitor a head start. Besides, when a big company wants to give you a hard time, they'll tell you to go to hell with your patent and send in their lawyers with pitchforks to speed you on your way down."

Of course, Bill was right in terms of many of the pharmaceutical and cosmetic products he had worked on during his big-company days. They were what I call "eminently substitutable" inventions: You change an ingredient or a processing step and you're around the patent.

But he was wrong with respect to his chewable tablet. Working together with our patent attorney, we identified a core invention in this tablet: When normal pharmaceutical tablets are stamped out in a press, they are hardest at their equi-center because the force comes from all sides and meets at that point. The Valentine tablet system did the opposite: It created a nice hard "shell" on the outside and a relatively porous center, which explains why it turned into a creamy liquid when it was chewed and exposed to the saliva of the mouth. The fact that most people and many doctors consider liquid medication superior to tablets gave Bill's invention market value as well as novelty.

This was a fundamental, observable difference. It is the kind of difference that makes both good patents and sellable inventions. It pays to look at your invention again and again to find a plainly discernible difference—and then to express that difference in your claims.

Several years later, after Bill and his son-in-law, Don McDaniel, together with Bill's financial backers, Bob and Bill Adams, formed a joint venture with PRI, and after we had obtained a carefully crafted patent on the tablet system, we sold field-of-use licenses to several major pharmaceutical companies in the United States and abroad. One of these companies, Marion Laboratories, Inc., bought out its license for calcium tablets by handing our joint venture a check for just about a million dollars. The settlement made sense for Marion because it discounted projected future royalties. It made sense for us because it provided needed working capital for further developments.

But it would never have happened at all without a clean-cut, issued patent. That patent gave Marion's management tangible justification for the payment. And it gave Bill Valentine a lot of pleasure.

Another and rather extreme example of the value of patents has to do with what I call the "nothing patent." Some years ago a man named Robert Baron came to me with an aerosol spray that would keep a woman's makeup in place all day long, even in extreme heat. It really worked so I was amazed when I read his patent—because the product had nothing in it. The explanation was simple: The aerosol's propellant served to harden the makeup. The product was called Lasting Beauty and I licensed it to Stuart Hensley, then president of Warner-Lambert. Baron made a nice sum of money while the product was on the market—it didn't last long—and it launched him on the road to financing and promoting technologies, including the process for colorizing black-and-white motion pictures. Sometimes, even a "nothing patent" is better than no patent.

In sum, whether you are an independent inventor, an investor in technologies, or a corporation executive, knowing how and when to take advantage of the patent system is too important to be left to the lawyers.

A Patent Primer

You don't have to be a lawyer to understand the essentials of patents. But if you take time to learn the basics, you will be a much better client. In the final analysis, you the inventor or the invention owner have to decide what are the keys to discouraging competition and then make sure that your claims reflect these points.

Just for the record, a U.S. patent, granted by the Patent and Trademark Office, an agency of the Department of Commerce, gives the patent owner the exclusive right to make, have made, use, and sell for a period of seventeen years the products or processes covered by the patent claims.

There are many good books on patents, how to draft them, file them, prosecute them. In the section An Inventor's Bookshop at the end of this book, I've listed eleven good sources under "Information on Patenting" and "Protecting Intellectual Property." If you want a fast education on the fundamentals, spend an hour or so with Hoyt Barber's 1990 book, *Copyrights, Patents & Trademarks*. It's also useful to know patent lingo. There's a helpful short glossary at the end of this chapter.

What follows are some practical insights on ways to make patents work for you and to frustrate the opposition.

The Business Side of Patents

Seventeen years' exclusivity—great! Provided, that is, that someone doesn't come along and pay the $1,500 fee required for a reexamination of your patent, saying it was granted improperly. Provided somebody doesn't sue you, saying that your patent partly infringes a prior patent of theirs. Provided someone doesn't go ahead and make a product like yours and tell you to go fly a kite.

Then, why go for a patent at all?

The U.S. patent system was started back in 1790, but unlike King George's grant of land to William Penn in 1681, the grant of a patent is not enforced by a king's army but rather by an army of lawyers, judges, and expert witnesses. To protect yourself, always remember that a patent is simply a business tool, not an intellectual suit of armor. You have to manipulate that tool for yourself.

Your patent lawyer—if he or she is a good lawyer and forthcoming—will point out the vicissitudes of the patent system and protect you against technical errors. But if you're going to get rich by using patents, you must recognize that many of the actions on patent matters are business decisions, not legal judgments. For example, do you want to file patent applications on every improvement you have discovered since the original filing or do you want to keep some of them as trade secrets until competition rears its ugly head and then spring your improved product as a surprise? Another typical business call is whether and when to go after infringers and whether to hassle their customers as well.

As a famous CEO once told his lawyer who was being too conclusionary at a board meeting: "You just tell me the risk and the rap. I'll make the decisions."

Before You Go for a Patent

When you first see an obviously worthy invention, your reaction as the inventor or as business manager is likely to be, "Patent it!" That's not exactly the right reaction. Your first thought should be, "Let's find out if it has been done before. We need not only a patent search but a literature search. We'll discreetly check with knowledgeable people in the industry and in academia. And concurrently we'll investigate all the other ways we can protect this invention."

Just because you haven't seen it in the marketplace doesn't mean that there isn't a patent lying in the files that anticipates—or interferes with—your invention; after all, less than 20 percent of the patents granted get commercialized.*

The point is, a strong patent is invaluable and a weak patent is

*There are some informed sources who put the number as low as 5 percent.

often an invitation to infringement. So you want the best possible preview of your prospects.

There's a wide variation in patent searches depending on who does them. They vary in price and in thoroughness. There's also some merit in doing the search yourself. As Dan Friel, inventor of the Chef's Choice knife sharpener, told me,

> Working through the shoe boxes in the disorganized Patent Office can take days or possibly weeks to be reasonably certain whether your idea is novel. Don't depend upon formal searches by an attorney or a paralegal and don't depend on the Patent Office once you've filed an application. The Patent Office itself usually does only a superficial search. It's a hollow exercise to obtain a formal patent without being sure whether the patent is in fact any good or enforceable.

He's right, of course, on most counts, but it takes experience to know your way around those patent files. Furthermore, few inventions call for a personal patent perusal as the first step, especially since you have to do that search at the Arlington, Virginia, Patent Office headquarters; any search at the regional patent libraries, whether in New York, Philadelphia, San Francisco, or Chicago, will be educational but hopelessly incomplete.

My belief is that you should search in stages: First, order a simple, low-cost search to be done by a quality professional searcher at the Patent Office. If that doesn't turn up a conflicting patent, then have the invention screened through a computerized data bank of U.S. patents, such as Lexis in the United States or Derwent in England. At the same time, do an electronic data bank literature search to see if there has been public disclosure of the idea by a third party. If the invention still looks clear, have an international check run to see that the invention hasn't been anticipated elsewhere in this shrinking world.

Bear in mind that all of the foregoing covers only a novelty search to determine whether this technology and the purpose to which it is put is new to the world. It gives you no assurance that your invention will not infringe some existing patent that is not overtly aimed at the product you envision. So you have to follow your novelty search with an infringement search, which is more expensive but extremely valuable, especially if you want to license the invention to a major company.

How much time and money you spend searching is a business

call. When a big opportunity is involved or when major money must be spent before reaching the market or if you expect to collect major money by licensing your patent, it's absolutely essential to be as thorough as possible. Failure to search every possible avenue has led to big liabilities for companies that have marketed products successfully, only to be sued by obscure patent holders.

It isn't as simple as just looking for products on directly equivalent inventions. Mike Lyden, senior vice-president of Tyco Industries, pointed out the problem to me:

> People invent things that probably will never work in the embodiments they envision. And they patent it even though they realize it doesn't work. Then someday someone comes along and makes changes so it really does work, and the patent holder says, "I'm in fat city," and he sues.
>
> We had that happen with our development of the Hovercraft toy. We did a patent check and found nothing that would seem to interfere. After a very successful introduction, we were contacted by a gentleman who claimed to have a patent on the device. Actually, his patent was for a much larger device intended to move products in a warehouse using a hovercraft design. His invention would never work for a toy product, but still he expected to be compensated, and, in the end, who knows, perhaps he will be.

Always, even if you do the initial search yourself, do serious searches through your patent lawyer. It's true that law firms add 30 to 50 percent to the basic cost, but a good patent savant will also add his or her analysis of potentially conflicting patents, and that's worth much more than the markup.

Making Your Patent Potent

The disturbing truth is that you can always get a patent—that is, if you're willing to compromise what you claim.

For example, patent examiners love to narrow the claims or to insist on so many qualifying clauses in the claims that the patent becomes a weak reed. And some patent lawyers acquiesce on the grounds that it becomes easier to distinguish your patent from the next one and thus get the application approved. But this also makes it easier for a competitor to get around your exclusivity. My advice is to fight as long as you can in defense of your claims short of total rejection.

As a money-minded businessperson, you should always take a marketing approach to patents. Picture the end product and work your way back to a patent that will protect it.

Getting it right applies not only to the words but to the drawings. When Alexander Graham Bell prepared his application for a patent on the telephone, he hired Lewis Latimer, an inventor and engineer who truly understood Bell's discovery, to draft the drawings. Latimer's very explicit drawings helped Bell win approval of an invention about which the examiners had little prior knowledge. Even then, you see, it paid to reach out and touch the right someone.

For your patent to be an effective business tool, it has to be written in such a way that the claims block others from doing what you're doing or anything close to it. Ask yourself what you would do if you were a competitor trying to invent around this patent. If you can't write a claim that clearly blocks potential competitors, at least try to write it so that it doesn't help them. And if you can't do that, forget about a patent and go for one or more of the other proprietary picket fences.

Now all that may sound obvious, but implementing it takes a lot of thinking. When you write your claims, picture them being read not only by the patent examiner but by a covetous competitor's attorney. Make competitors think twice before infringing. Try to envision how a federal judge might someday interpret the claim if you are forced to sue a big corporation for infringement.

It isn't enough that *you* understand the meaning and implications of the claim; the world must understand it too. And, as I learned time and again, if you have only one patent opportunity at hand, start looking immediately for a patentable improvement. And start building a strong technology package as well.

In sum, some inventors have a talent for writing clear, preemptive claims; others don't. Some patent lawyers have it, some don't. So both the patent lawyer and the inventor should work at writing the claims, and it will help if a businessperson with a marketing viewpoint is looking over their shoulders.

What Is Patentable?

From a businessperson's viewpoint, it's worth remembering the many different kinds of things that can be patented. Section 101 of the patent law is very broad:

> *Inventions Patentable.* Whoever invents or discovers any new and useful process, machine, manufacture, or composition of matter, or any new and useful improvement thereof, may obtain a patent.

While utility patents, which most often cover the physical or chemical properties of a product or its production process, are by far a stronger line of defense, design patents can also be useful. Their limitation, of course, is that they cover only the one particular design specified or a "colorable imitation," meaning not an exact copy but close enough. This makes design patents only as good as the marketing effort behind them, which in turn means that, to be useful, the specific shape and appearance of the product must identify "the real thing." For example, wouldn't you like to own a design patent on the Coca-Cola bottle?

All in all, a design patent, especially if added to a utility patent portfolio, can be an important impediment to would-be infringers.

You can also patent plants, you can patent a new use for old materials, you can patent animals, and so on. The point is, whatever new business or new product you are launching, take a good look at the patent possibilities. A well-thought-out patent, issued or pending, is useful in initial marketing, and it could help keep the competition at bay until you've established a market position.

A Picket Fence of Patents

I am often asked about going after multiple patents, especially by those who are cost-conscious. The fact is that for several reasons a picket fence of patents is often geometrically more valuable than a single patent.

First, they slow down the other guy's lawyers. Opposing patent counsel knows that, realistically, if *any* of the claims in *any* of the patents stand up in court, his side loses—and this deters him from chuckling to his boss or client, "Sir, I'm reasonably confident we can win this one."

Second, the inventor and his patent counsel can use multiple patents to make wonderfully clear the various applications of a basic discovery.

Finally, because the applications will probably be filed sequentially, they can extend the date of the last-to-expire patent, unless otherwise agreed with the patent examiner.

One of the finest examples I know of a company using multiple patents to the nth degree is Lanxide Corporation. Lanxide is an emerging growth company that creates new kinds of "engineered materials" by combining ceramics and aluminum under extremely high temperature and pressure. The Lanxide technology provides several different advantages: erosion-resistance, impact-resistance, heat-resistance, dimensional stability—plus the ability to manufacture complicated shapes to very close tolerances directly from the material without machining.

Marc Newkirk, founder and CEO of Lanxide, told me,

> I believed from the beginning that this process would create a whole new forms of matter. Then when we started to test the applications, it became clear that we couldn't even forecast how many fields of use we would enter—or how many new discoveries we would have as we went along.
>
> So, from the beginning, we set aside an unusually large percentage of our capital for patents. Bear in mind that this capital came from private investors and from joint venture partners like Du Pont and Alcan—people who wanted to see real income-producing products as soon as possible. But we knew that in the long run, if we were to become an industrial giant, we would have to protect our position at every turn.
>
> As of the end of 1989, we have filed nearly 2,300 patents worldwide. In 1989 alone, we filed 755 domestic and foreign applications and received notice of allowance for 49 U.S. and 72 foreign patents. This patent portfolio not only provides protection on specific aspects of the technology and its applications, but it also is a powerful force now in persuading potential customers to place initial orders for Lanxide parts. *And* you can be sure it was an important factor in raising $10.8 million of additional capital last year.

Elsewhere in this book, you'll read more about how Marc Newkirk put Lanxide together from a flash of insight, how he grew it from an idea to a company in which the 1991 value of his own stock and options was several million dollars, even before the company went public. Although you may not want the kind of patent bill that Lanxide faces, you should always consider the value of going for more than one patent, and, whenever possible, keep at least one application pending to worry the opposition as to what you will claim next.

The Right Patent Lawyer

How do you pick a patent lawyer who will get your claims just right? It's a matter of matching the specifics of your patent problem with the experience and attitudes of your chosen counsel.

When Leonard Chavkin and Leonard Mackles came to me with their invention for dispensing pharmaceuticals via aerosol, producing a soft whip (almost like whipped cream) rather than a liquid, it was easy to see the advantages: no spilling, better taste, no open bottle, and so on. They had been working with a patent lawyer of scholarly bent and international experience. However, when I read the initial patent application that he had drafted, I told my partners at PRI, "We have a problem here. This application describes very well what the invention is. But we need a patent application that portrays why the products we foresee could never be made without this technology. We must not only convince the examiner to grant our claims but also impress licensees' lawyers and R&D people who may want to give us a hard time."

I took that patent application and a product sample and the inventors' original disclosure to Al Engelberg, then a partner in the patent law firm of Amster, Rothstein & Engelberg. First of all, Al was trained in the chemistry of pharmaceuticals. Second, he had spent his early years working as an examiner in the Patent Office. Third, and most important, he represented the Generic Drug Manufacturers Association and some of the generic companies as well. Al had spent a lot of his time studying drug patent applications and how to break them.

I figured, if he knew how to tear patents apart, he ought to know how to write the kind of patents that would make company lawyers say, "Hey, boss, we better take a license if we want to use this invention."

And that's exactly how it worked out. In January 1987, the patent was issued. The claims were broad and basic; the examples cited were specific to major product categories. Half a dozen big companies, in the United States and abroad, licensed the aerosol whip. They had the patent reviewed by both inside and outside counsel, but not one ever challenged the patent position. Sadly, the first products made with that wonderful technology encountered clogging, but the patent covers such a wide spectrum of products that it should eventually prove quite valuable. In the meantime it has thrown off well over a million dollars in option and license fees.

Writing the Patent Application

Here are some practical guidelines to help you select the right people to draw up your patent applications, and especially the claims:

- Pick a patent lawyer who not only has technical training in your field of invention but is knowledgeable about the marketplace for such products. The Martindale-Hubbell directory, available in many libraries, will tell you the area of training of every patent lawyer: chemistry, electrical engineering, mechanical engineering, and so on. You can also check the directory published by the American Intellectual Property Law Association. But neither directory will tell you a lawyer's business experience or negotiating skills. You have to probe for these or check with his or her other clients.
- Make sure that your lawyer both understands what product or products you expect to protect with this patent and reflects this understanding in all his correspondence with the Patent Office, thus creating a patent file that will help your case in the event of a lawsuit.
- Pick a lawyer who has patent litigation experience, or at least one who has a partner with such experience, to review the application before it is filed. The time to start planning your patent defense is before you file the application. One useful technique is to ask your own team to "invent" a product or method that might circumvent your patent claims and then ask your attorney whether or not they have created an infringement or a reason to revise your patent strategy.
- Ask someone with marketing expertise to look at the invention and tell you what the product "headlines" might be, that is, how you might express what the invention will *do* for people. Then communicate this to your patent lawyer, asking him or her to get the patent claims in harmony with the product news. If you're seeking to license a corporation, this helps because management can relate the protection they'll be getting to the product they'll be launching. Harmonizing the product news with the patent claims also has a sobering effect on would-be competitors and, if push comes to shove, on juries.
- If you or your attorney should meet with the patent examiner in an effort to get your claims approved, let the examiner know the

various kinds of products to which your invention applies. Ask the examiner for comments as to how well those claims cover the kind of products you have in mind. Whether the examiner's reactions are on or off the record, play them back to him in any revisions of your patent filing.

Suing and Getting Sued

Voltaire once said philosophically, "I was ruined only twice: once when I lost a lawsuit and once when I won one."

The history of many inventors lends credibility to Voltaire's whimsy. Charles Goodyear, discoverer of the vulcanizing process for Indian rubber, holder of U.S. Patent No. 3,633, instead of becoming a wealthy man for having revolutionized the rubber industry, lost control of his exclusive rights—primarily by granting badly written, incomplete licenses to others. At last, in frustration, for the sum of $15,000, he hired the great orator Senator Daniel Webster to represent him in court. The bad news: Goodyear kept losing many cases. The really bad news: When he did win, the average award was under $100.

Today, patent lawsuits have a heavy cost in money, time, and emotional energy. The best way to resolve a patent dispute is to settle it out of court. If you doubt this, ask Robert F. Kearns, who sued the Ford Motor Company over his intermittent windshield wiper invention; he turned down a $30 million settlement offer and pursued the case to victory, but after twelve years in the courts, in 1990 the jury awarded him only $6.3 million plus interest, which could double that amount. What Kearns finally put in the bank after legal fees and costs might seem a princely sum to many; but to this long-frustrated inventor, it seemed paltry.

An interesting sidelight: Because Kearns did not file his lawsuit until years after the infringement had begun, the jury awarded damages for only ten of the sixteen years the wiper system was in use.

Of course, Kearns still has cases pending against General Motors and other automobile manufacturers, who are likely to settle in light of the Ford case.

The moral once again: Sue if you must, but settle if you can! On the other hand, remember that the best way to get a decent settlement is to be—or at least to appear to be—ready to go to court and through the appeals system.

Since the establishment in Washington in 1982 of the Court of

Appeals for the Federal Circuit ("the patent court"), and since the rehabilitation of free enterprise and entrepreneurship in the 1980s, there has been a tremendous change in how the courts and juries treat patents.

In the 1970s—and especially in some of the southern courts where there was something of a "poor boy country judge versus you city slicker lawyers" attitude—a large majority of the patents challenged in court were declared invalid. During the 1980s, this situation was reversed. If the patent was well conceived and well written, it usually got respect in the lower courts, which had to live under the threat of reversal by the patent court.

Moreover, in cases of willful infringement, the patent court has shown a willingness to award attorney's fees, to multiply damages, and to allow permanent injunctions to stand while appeals are pending. Double and even triple damages have been awarded where infringement was blatant, and in special cases the patent court has even imposed personal liability on corporate officers who recklessly authorized willful infringement.

In Polaroid's suit against Eastman Kodak, it is noteworthy that the patent court backed the district court's decision to let a permanent injunction stand pending the appeal. This effectively put Kodak out of the instant picture business and sent a chilling message to companies that disrespect for patents can be perilous.

It also led ultimately to a $900-million award to Polaroid. Please note, however, that while this is a very large sum of money, it was substantially less than the $1.5 billion to $2 billion that most observers were predicting.

Why did the court award so much less? One important reason, as one of Kodak's defense lawyers for this phase of the trial, Robert Frank, cogently argued, was that Kodak had consistently over the years asked its patent counsel, the famous Kenyon & Kenyon, for written opinions as to whether it was infringing the Polaroid patents and whether or not certain Polaroid patents were valid. The judge concluded that, although Kenyon's opinions were wrong, they had been closely reasoned and it was therefore not unreasonable for Kodak to have relied on them. And when Polaroid's counsel claimed that Kodak had committed willful infringement, the judge turned away the argument with the famous phrase, "That dog will not hunt." Both sides threatened to go to appeal but settled in 1991 for $873 million plus $52 million in interest, a total of $925 million, a record-breaking patent infringement award. Meanwhile, respect for patents was reinforced in the boardrooms of corporate America.

All in all, for both large companies and small, the new judicial appreciation of the role of patents in a capitalist society has encouraged inventors to invent—and to sue when necessary.

A New World for the Little Guy

In the late 1980s juries began awarding headline-making sums to independent inventors who had sued giant corporations.

In Chicago, after a twelve-year battle, a federal court jury directed Mattel, Inc., to pay $24.8 million, plus almost twice that amount in interest, to Jerry Lemelson on the grounds that Mattel's Hot Wheels toy infringed Lemelson's patent on flexible plastic tracks. Furthermore, it was ruled that because Mattel had willfully infringed the patent, the judge could award triple damages. While it is conceivable that the entire verdict could be reduced or even overturned on appeal, the Lemelson-Mattel case—and other recent verdicts—made headlines that sent some chills down the spines of corporate managers whose policy had been to tough it out rather than settle.

Now Jerry Lemelson is a brilliant inventor with some 450 patents to his credit. But he is not everybody's favorite inventor. Some people allege that he files patent applications less with the idea of commercializing the product than for the purpose of harassing some manufacturer. Nevertheless, independent inventors can learn a lot from Jerry Lemelson, and company executives can study him as a cautionary tale.

Mattel isn't the only company Lemelson has sued. He has been involved in more than twenty patent lawsuits, and he has won less than half of them. Early in his career, he sued a number of toy companies and although he lost against Fisher-Price and Ideal Toy, he won enough times to make a very good living. As his career matured, Lemelson created more sophisticated inventions, one of which led to a suit against IBM, and Lemelson collected from the giant.

Not only is Lemelson enormously prolific, he often creates his millions strictly from his mind, not from tinkering. According to a lawyer whose firm has been on the opposite side of the courtroom in Lemelson cases, Jerry seldom builds a working prototype of his inventions. Said this attorney, "Lemelson is so smart about writing patents that he could have gotten a patent on the airplane in 1862, time out of mind."

No doubt, Jerry's a guy who understands structuring patents in ways that many of us could emulate. He conceptualizes the end product and then he himself writes the patent application specifically to fit the concept. In other words, he has made a business of patents.

If more companies thought in his terms, they'd double the number of marketable products produced by R&D.

Another remarkable example of a man who knows how to manage invention is Gene Lang, president of Refac Technology Development Corporation. He is best known to the world for an idea that he is happy to have people copy. He promised a class of sixth graders in a Harlem school that if they finished high school, he would pay for their college education, each and every one of them. He is also known for having helped Gordon Gould obtain his basic patents on lasers, a story detailed in Chapter 6.

What is less well known, except to the targets of Refac's lawsuits, is that Gene and his late associate, Philip Sperber, acquired rights to certain patents of independent inventors, patents that affected major product categories, and then launched infringement suits against the manufacturers of such products, their distributors, and retail customers—frequently all at once. The product categories range from electronic keyboards to liquid crystal displays to video recorders to audiocassettes to spreadsheet software, including Lotus 1-2-3®.

One of the most spectacular suits was filed by Refac in 1988 against a dozen major corporations, including Federal Express, United Parcel Service, Sumitomo Electric, Eastman Kodak, Honeywell, and UNISYS, claiming that they were infringing Refac's patent on bar codes.

All in all, Refac is reported to have participated in over 2,000 lawsuits. After the passing of Philip Sperber, however, Gene Lang was apparently less and less comfortable with the onslaught that had been launched by his company. And when the Federal Appeals Court dismissed many of Refac's actions and verbally slapped the company's use of lawsuits as a bargaining tool, Lang publicly backed away from the "sue 'em all" philosophy.

Nevertheless, many dollars were collected from companies that settled, and Refac's revenues from licenses and settlements exceeded $10 million in 1989. Meantime, Refac's interest in the Gould laser patents—about 16 percent—will swell Refac's income stream for years to come.

Rarely a day goes by without a report in the business press that somebody is suing somebody else on a patent matter. Small companies sue large ones; large companies sue everyone. The biotechnology companies may be at it forever.

And note too that the power granted by a patent is not absolute. A relatively small Florida company, Windmere, won a multimillion-dollar award against North American Philips on the grounds that Philips had violated antitrust laws in its business behavior. The court ruled that, even though Philips enjoyed some degree of monopoly by virtue of its patent on rotary electric shavers, that particular government grant did not supersede the government's antimonopoly laws.

Even the giants of the pharmaceutical industry are not immune to attack on their patents by smallish companies. As this paragraph is written, Canadian-based Genpharm, Inc., one of a group of generic drugmakers controlled by the Tabatznik family, is seeking to invalidate mighty Glaxo Holdings's second patent on Zantac, its $2.5-billion antiulcer product, on the grounds that this second patent is really only a variation of the first patent, which expires in 1995. Stand by.

A Stance for Independents

In short, you have to be prepared to be litigious because that's the way the business world works today, especially if you're successful. But I repeat, patents are a business tool, not a permanent entitlement like appointment to the House of Lords. So be prepared to fight but always look for a way to settle. Back the other guy into a corner—where there's a door.

Suing people, especially big corporations, costs a lot of money. However, despite what many "uptown" lawyers will tell you, some top-drawer attorneys will take cases on a contingency or partial contingency basis. Lemelson's Mattel case was widely reported to have been handled on a percentage-of-results basis by Jerry Hosier of a prestigious Chicago law firm. Robert W. Kearns, whose landmark victory over the Ford Motor Co. on the intermittent windshield wiper will make him a few million dollars, was represented by Arnold White & Durkee, a very large and respected patent law firm headquartered in Dallas, Texas. They, too, it was reported, worked on a share-of-the-action basis.

Infringement cases aren't the only kind of legal work for which patent lawyers will make imaginative financial arrangements with their clients. I know one lawyer who successfully prosecuted a suit for his client without any fee or guarantee. That suit resulted in the invalidation of a competing company's patent and allowed his client to market a multimillion-dollar product almost overnight. That attorney's compensation was based on a substantial percentage of the added profits that his client realized from the product.

While some patent lawyers will handle patent and trademark work for a percentage of the royalties or at least on a deferred-fee basis, other attorneys decry this practice as unprofessional and, *de minimis,* untraditional. However, from a strictly business viewpoint, flexible charging arrangements make sense for lawyers only when the licensing potential is great or when a start-up company looks like a long-term client or an infringement case has the potential for generating major awards. An open discussion with your attorney on the value of your patent(s) and/or the possible recoveries from infringement suits is a useful prelude to any discussion of fees.

Lou Heidelberger, who heads the TechLex Division of Reed, Smith, Shaw and McClay in Philadelphia, speaks for many attorneys when he says, "Like most law firms, we fill most of our sixty-hour weeks with billable time for present clients. But when we see an invention or a start-up company that has the makings of a substantial future client, we start thinking creatively about a financial arrangement that will make us an integral part of that situation for years to come."

If, however, you have deep pockets and can afford to hire counsel on a fixed-fee or hourly basis, do so. Why give away part of the action when you don't even know how big it will be? But if you're still at the struggle stage or if your company's running on a tight budget, don't hesitate to ask your lawyer for financial cooperation.

From the cases presented here and many others, both recently decided and pending, you can see two major turning points for inventors and corporate executives:

1. The judicial interpretation of patent law, and the climate in which legal actions are decided, has swung in favor of the patent holder.

2. High-quality legal counsel, skilled not only in filing and prosecuting patents but in their enforcement, are available to almost every patent holder who has a good cause.

Obviously, this is good news for inventors and smaller companies founded upon technology. In the long run, it is also good news for larger companies because respect for patents per se translates into longer product life and improved profit margins for companies with powerful marketing forces, whether the patent is generated by internal research or licensed from outside.

Deciding on Going to Court

How does one decide whether or not to invest years of effort and financial sacrifice in fighting for one's rights in a court of law? The answer is: very thoughtfully.

Most of that thought has to be devoted to a critical, ultra-hard-nosed analysis of the facts. What do the patent claims actually cover when you read them without the glow of righteousness that comes with being the patent holder? How would you skirt around them if you were hired to do so and how close is the alternative to what the patent teaches? What does the paper trail indicate? If you had prior discussions with the alleged infringer, what kind of confidentiality agreement was signed?

Look at every scrap of the record. If you do not find any obvious weakness in your case, sue—but be prepared to settle. If you do find an Achilles' heel in your case, threaten to sue, then try to settle as quickly as possible.

Losing and Winning and Why

Naturally, no two cases are alike. But let me cite several legal entanglements of particular interest because they all involve the Johnson & Johnson "Family of Companies."

I'm not really picking on Johnson & Johnson—there are many equally prominent exemplars of the problem—but to America as a whole, J&J is viewed with only slightly less reverence than Motherhood. It starts as we curry our kids with Johnson's Baby Powder and Baby Oil and continues through J&J's statesmanship in dealing with the Tylenol murders in Chicago.

However, lest you have to discover it for yourself, this is a reminder that most corporate hierarchies, including J&J's, are made up of forceful businesspeople who are extremely reluctant to pay ten cents in royalties to anyone—if their lawyers or R&D people think they can escape doing so. It's a rule: When choosing a company to license or sell to, consider how litigious they have been in recent times.

J&J was sued by Minnesota Mining & Manufacturing Company for willfully violating 3M's patents on fiberglass tape used for medical casts. 3M won a judgment of $116.3 million. J&J sued the Food and Drug Administration to stop U.S. Surgical Corp. from entering the $1.3-billion surgical suture market. J&J's motion was denied.

Over the years J&J has also been sued for patent infringement by a number of inventors. One of them was Dr. Raymond V. Damadian, the father of magnetic resonance imaging (MRI), a major scientific breakthrough. Another was Vincent R. Sneider, a high school graduate who invented and patented a panty pad for feminine hygiene.

A physician whose field was fundamental research in biophysics and biochemistry, Ray Damadian—driven by the memory of his grandmother's painful, lingering death from cancer—had centered his attention on how the living cell generated electricity. He linked together two hypotheses: (1) that a healthy cell must be different from a cancerous cell as to structured water, and (2) that a nuclear magnetic resonance ought to be able to detect the difference. (The common terms for such machines today is magnetic resonance imaging, or MRI, the word *nuclear* having proved entirely too ominous for doctors and patients alike.)

After proving his theory by experiments, Ray took seven years, all the while scrounging for grants, loans, and private investments, to build his first MRI scanner. He then formed Fonar Corporation in 1978 to build and market the equipment, and went public in 1981. At that point, the Food and Drug Administration slowed down the parade by not issuing marketing permissions until 1984.

By then, Johnson & Johnson and General Electric and some Japanese companies were racing down the road to market MRIs. In 1982, Damadian sued J&J in New York State, where Fonar is located. As the defendant, J&J promptly had the case moved to the federal court in Boston. Because J&J is headquartered in New Jersey and Fonar in New York, the reason for this switch in locale is obscure, but, according to Damadian, J&J frequently chose this venue.

At the end of a seven-week trial, the jury held that Fonar's patent was valid and infringed, but the presiding judge reversed the decision of the jury. That's called Judgment Notwithstanding Verdict. The decision was appealed. Fonar lost again. Damadian tried to get to the Supreme Court of the United States, but it would not hear the case. Then, after spending $2.2 million in legal fees, he gave up the fight.

Now, not to defend Damadian or Johnson & Johnson or the judge, but purely to stimulate your appreciation of how much patent claims matter, please "listen in" on excerpts from my conversation with Ray about his lawsuit:

Alan: If you could do it over again—you know, when you get a few gray hairs—what would you say to the next generation so that they could do it a little better? Would you write the patent differently?

Ray: With perfect hindsight, I would have done a lot of things better, including the patent claims. You know I sat through a seven-week trial. But I don't think there's any way I could have written it differently in 1979.

Alan: With your present knowledge of the patent office and the patent system, would you now *think* differently in writing such a patent?

Ray: Everything that was written, and the limitations on what was written, resulted from my vision of what was going to happen. I just didn't see all these developments back in '72. Somebody should have seen it, but I didn't. I was in the best position to see it and I just didn't.

Alan: What were the key issues on which they found holes to poke at you?

Ray: Well, they really didn't. The judge shouldn't have done what he did. If they had honored the claims the way the claims were written, the patent should have held.

Alan: What court were you in?

Ray: Boston. And we won before the jury. The jury held the patent to be valid and infringed, but the presiding judge who sat on the case six weeks later reversed it. Now the basis for his reversing it was that my original conceptualization of this consisted of a body scanner that would hunt for these signals and it didn't specify whether you'd make an image from the signals or whether you'd write the data down—it just didn't deal with that. And what the claims said—they were broad—they didn't specify whether you were going to do it in a human body or whether you could do it on a biopsy. In that sense, I think the patent was done properly... it was general. My second embodiment dealt with building the actual scanner itself with a man standing inside of it and the probe going back and forth on him, and all it said was, compare the measurements of the signal in cancer with the measurements of the signal in a normal case and you can detect whether or not there is a cancer based on that.

The opposition argued, well, these are not images. We argued that you can't make the scanner, you can't make images without this data. It's the data that make the distinction. And then the judge just changed it, he got it all fouled up, you know.

There was one argument in the courtroom that turned out to be true after my original patent issued: that the signal you got from cancer tissue was not necessarily different from the signal you got from other types of diseased tissue. Consequently, on the signal alone, you could confuse the cancer tissue with other types of diseased tissue. Well, that's only one element of the invention. That means that we can't yet make a positive diagnosis based upon the uniqueness of the signal, but if the signal is different from normal—which it was—we could detect all diseased tissue, which is what the scanners are doing today. They see these white spots in the brain and those white spots in the kidney and in the liver, and those white spots are different from the surrounding tissue, so when you picked up this abnormality, you would put a needle in there and do a biopsy. You'd be doing what the X-ray scanner can do, but infinitely better.

It's fundamentally dependent upon the fact that the signal from the cancer tissue is reliably different from the signal from the surrounding normal tissue.

The claim covered distinguishing cancer tissue from normal tissue of the same type. The judge said, you can't distinguish cancer tissue for any other diseased tissue. The six-person jury didn't have any trouble with that point. They wrote notes the whole time, and the the judge advised them to do so. And they very astutely dissected out some claims that they felt didn't hold and other claims that they felt did.

Alan: Then, if you had specified "diseased tissue" instead of "cancerous tissue," the judge could have lived with that?

Ray: Well, I think there were other motivations going on with him. I'm not comfortable about speculating. There was a lot of money at stake and he was hostile from the beginning.

You see, when it comes to enforcing infringement, you don't go to the patent court, you sue in the federal court and that's a serious problem. Because many of these federal judges don't know anything at all about patents. In my case, the tragedy is, now I've got to go out there and do battle not only with J&J and GE but with the Japanese companies that are coming to push my own invention out of the domestic market.

Alan: Did you go for additional patents as you went down the line with your invention—say, on the conversion of the signals into images or such things?

Ray: Oh yeah, I think I have about seven or eight other patents and we're generating others all the time.

Alan: I've heard many points of view on the usefulness of the patent system. At this point, what do you think?

Ray: I think they don't work. I think they're just a silly business.

You've been reading the heartfelt and rather emotional words of a man who spent seven years of his life in an underequipped laboratory, often working eighteen hours a day to solve the mystery of detecting cancer and identifying its locale noninvasively. Small wonder that he felt the world should recognize the validity of his discovery and honor the claims of his patent.

Fonar Corporation in 1989 did about $60 million in sales and made about $2.6 million in profit. Ray Damadian was elected to the National Inventors Hall of Fame, alongside such giants as Thomas A. Edison, Alexander Graham Bell, the Wright Brothers, Eli Whitney, Samuel Morse, George Eastman, and Edwin Land. But as he said to me wistfully, "Sure, I've made a few million dollars, but if I'd won that lawsuit, every one of the MRI makers would have taken a license from me and I'd be one of the richest men in the country!"

Whether or not Raymond Damadian's patent deserved to be as basic to MRI as Gordon Gould's patents appear to be for lasers lies in the never-to-be-relived past. The image that resonates from Ray's experience, though, is sharp and clear: Claims that win in court must be specific and complete, and certainly the dependent claims, or CIPs (continuations-in-part), must go to the heart of the practical applications, even if the basic claim is broad and fundamental.

In striking contrast to Ray Damadian is Vince Sneider, an independent inventor with a high school education, a fertile mind for invention, and the disposition of a bulldog toward a world he considers out to harm him.

Vince is also a very good golfer—which accounts for how he became interested in sanitary napkins in the first place. Often invited to elegant clubs as a "ringer," he found himself playing one day in a foursome with Mickey Mantle and two gynecologists from New Jersey. For whatever reason, the gynecologists started discussing the difficulty some women experience in containing their menstrual flow on "heavy days."

Sneider, who had fixed the furnace for the nuns at his high school at age 14, invented fast-repair techniques for the U.S. Air Force that kept planes flying during World War II, tinkered with electrical connectors, invented burglar alarms, brought dead automobiles to life, and so on, went home and asked his wife to explain the details of this menstrual problem. Within two days, Vince had the answer and built a model. Within weeks, he sold the product to the Kimberly-Clark Corporation.

A few months later, he turned his mind to the light end of a woman's period and designed a very thin pad especially for that time. Having obtained a patent on this invention, in 1973 he made major presentations to both Johnson & Johnson and Kimberly-Clark, makers respectively of J&J Sanitary Pads and Kotex®.

Both companies offered consulting fees while they tested the product against the prospect of a license later. Sneider decided to go with Kimberly-Clark, whose Kotex brand dominated the market. After several months of discussion, however, the Kimberly-Clark people told Sneider that they had decided not to go ahead with him. Not long thereafter, K-C's Lightdays product appeared on the market, and within a few months J&J launched Carefree, a similar product. To Sneider, these products looked for all the world like his invention. Obviously, J&J and Kimberly-Clark either felt that they hadn't infringed Sneider's patent or didn't think that this loner would sue.

Vince had a very tough decision to make. He knew that lawsuits take both money and time away from other income-producing activities. At the same time, the lawyers he consulted in Atlanta, where he lived, and in New York all advised him that he had a tough case, especially against these giants.

Two tragedies and one piece of good luck changed everything. In 1974, Vince's wife died, a suicide. The following year, his daughter was killed in an automobile accident. Understandably, his outlook on life changed. His bile boiling over, Vince sought out lawyers who would risk a difficult case, and, in 1978, he found a Chicago lawyer who accepted the challenge.

Within months, Johnson & Johnson made a six-figure settlement with Sneider, and, three years later, Kimberly-Clark did the same thing, making Sneider a millionaire.

"Why did you settle when your patent had years to run?" I asked Vince at his home in Atlanta. "Easy," he replied, "I wanted to get on with my other inventions."

Judgment Pending

When and how to go for a patent is as much a business judgment as it is a legal opinion. So is when to sue and when not to, and when to settle.

Looking back with 20/20 hindsight, Ray Damadian admitted, "I could have done a lot of things better, including the patent claims."

Vince Sneider never looks back. He's too busy inventing an answer to some new problem he has observed.

All this underscores the ultimate wisdom about patents: When you're going for the gold, first decide if you have something new and worth protecting, then hire the best minds you can find to write and review your application. Spend the time, money, and sheer effort that it takes to get it right—before the day comes when you have to stake your business life on it.

And, recognizing that justice under the law can be quixotic, seek as many additional kinds of protection as seem to make sense for your new product/invention/technology. Just read the next chapter and consider the possibilities.

Patent Lingo

Here are some common words and phrases you should know when working with patents and patent lawyers:

102*, Anticipation of the claim Rejection or invalidity of the claim based upon a single piece of prior art containing all the elements of the claim in the patent application.

103*, Obviousness of the claim Rejection or invalidity of a claim where differences between the invention and the prior art are such that the invention as a whole would have been obvious to one of ordinary skills in the art at the time of the invention.

Continuation A continuation application with another set of claims that you can still file even after the rejection of your patent.

Continuation-in-part (CIP) New or additional information you can attach to the original disclosure if you make an additional discovery for the same invention after filing the original patent.

Division A separation of one of the inventions from the original application if the examiner says your application really covers two inventions, and a filing of the other as a divisional application.

File wrapper The entire package of filings, correspondence, and notes relating to a given patent. Critical in the event of a lawsuit, the entire file is secret prior to issuance of the patent but available on demand from the U.S. Patent Office after the patent is issued.

Foreign filing Filing of your U.S. patent application in most foreign countries within one year of the U.S. filing date to preserve the original date of invention under the Paris Convention.

Office action Any response from the Patent Office.

PCT and EPA Single applications that permit multicountry filings. PCT is the Patent Cooperation Treaty and EPA is the European Patent Association. PCT filings cover most major countries of the world, and EPA applications cover most European countries.

Reissue application What you file for when you want to change the claims of or correct mistakes in an already issued patent.

*Section of the patent law referred to.

Chapter 5

Building Other Proprietary Picket Fences

In a utopian world, every creative person would have his or her own special niche, get a fair price for his or her innovations, and have time for a nap after a gourmet lunch.

But in a free-market society, it's more often a hot dog for lunch and dog-eat-dog business tactics.

In such a competitive world, it only makes sense to protect your profit margin in every way you can. Patents are a potent business weapon, but there are other tools and techniques for protecting the company jewels. Except for "black boxes," these other proprietary picket fences work synergistically with patents to entrench a product's position.

What's in a Name?

"Coke" or "Pepsi"—those are rememberable, easy-to-say brand names. "Royal Crown" is not. While there are many other factors that separate the Coca-Cola Company and Pepsico, Inc., from the Royal Crown Company, the natural zing of the brand names *Coke* and *Pepsi* is certainly fundamental. Can you imagine Michael Jackson singing a commercial with words like "Royal Crown turns me upside down"? Not exactly a natural. The only suitable tie-in for Royal Crown would seem to be Prince.

Now what makes a good trademark is a subject unto itself and is often hotly disputed. But choosing an appropriate and memorable trademark for an invention and then protecting it properly is a central task in marketing new technology-based products. The goal is to make your mark synonymous with the invention—but not too synonymous, lest it become generic for the category and be lost.

Trademark War Stories

In the United States, trademarks, once established in the mind of the public and the trade, are like the Dali painting *The Persistence of Memory:* They drape themselves over the landscape of the mind and have value for years and years to come.

Consider blue jeans. You might never have thought of blue jeans as an invention, but it was. In 1850, a man from Germany arrived in San Francisco intent on cashing in on the gold rush by making and selling tents and tops for covered wagons from heavy-duty cloth. When he heard miners complain that their pants frayed and wore through quickly as they panned for gold, he turned the heavy blue tent canvas into trousers, sewed and studded them securely, and developed America's standard garb for outdoor workers. The man's name was Levi Strauss and the product became known as Levi's, a trademark established by common law and later registered. Unfortunately, the company never could register the word *jeans,* which derived from the French word *genes,* a name for the cotton pants worn by Genoese sailors.

What made blue jeans an invention? Although Levi Strauss never applied for a use patent, his was apparently the first application to clothing of heavy denim, a stiff, relatively unworkable material . . . so it was new. It was functional, providing protection against abrasion with measurably improved wearing qualities . . . so it had utility. And it was not obvious, because the tenting material was not only prima facie unsuitable but also, in terms of its ability to be shaped to the human figure, not easily cut and sewed.

While today there are many makers of jeans, from unbranded cheapies to Ralph Lauren smoothies, Levi's remains the leader. "Levi's" are what you ask for when you want long-lasting pants for real work or simply want to make a social statement.

If there is such a thing as the perfect trademark, it could be PLAX. For an antiplaque mouthwash, you could hardly do better. It suggests what the product is used for without being purely descriptive—which would disqualify it as a trademark. It's short, so it can appear in large letters on the bottle and shout from the shelf, and it is easy to say and remember. It was certainly a major factor in the valuation of PLAX when Pfizer, Inc., bought it for over $200 million from Oral Research, Inc.

Another entry in the near-perfect trademark contest was Hidden Beauty® as a name for a padded brassiere. It avoided the disqualifying characteristic of being too descriptive, but it unmistakably projected the right image.

Unlike the value of patents, which is intrinsic, the value of trademarks usually depends upon astute and aggressive promotion by their owners.

Slap Wrap bracelets, a rage among children in 1990, became nationally known overnight by word-of-mouth publicity. But they were bedeviled by imitations from Taiwan. With the patent application still pending, the maker, Main Street Toy Company, had little protection except for the name that was slapped on the product itself with an iron-on label. Unfortunately, that kind of label wears off, or people take it off.

This became particularly important when it turned out that the imitation bracelets used steel strips of a lesser thickness, which resulted in broken bracelets that could and did cut children's arms.

So what do you do when the kid world goes crazy about your product, calls it by your brand name, Slap Wrap®, but buys a lot of the imitations that flood the market?

As Richard Murtha, an ex-Coleco official and a founder of Main Street Toy Company, explained it to me,

> The factories in Taiwan that jumped on the bandwagon ultimately did us a favor by making their product so bad. They used cheap steel and shipped by ocean transit. As a result, some of their products had started rusting by the time they arrived here. And their fabric tore. That made us take extra care to use not only stainless steel but certain very strong, resilient fabrics.
>
> When mothers reported the danger to consumer protection agencies, the result was a fiasco for the imitators but a PR disaster for us. We turned it into a PR triumph.
>
> We're a small company, but four of us are experienced sales and marketing people. For about a ten-day period, we just pounded out stories to every newspaper, every radio station, every TV station that we could find listed. We sent everyone interested a sample of our product showing the difference and followed up with phone calls. We did so many radio and newspaper interviews in one week in October

> that I stopped counting. I did count the TV interviews and there were nineteen. We had a nose for the fact that there was a fad rage sweeping the country—and made sure everybody knew that we were the good guys.

As a result of the war against the cheap knockoffs, Main Street Toy's Slap Wrap trademark received millions of dollars of publicity. The Slap Wrap bracelet is a novelty, a fad, as were the Hula-Hoop, the Pet Rock, Wally Wall Walker, and other skyrockets of hallowed memory. The challenge for Murtha and his associates is how to apply the Slap Wrap trademark to other products before the world considers it passé.

Trademark Asset Management

Once embedded in the public consciousness, a trademark may turn out to offer stronger protection than a patent. Consider Windex®. Marketed by the Drackett division of Bristol-Myers Squibb, Windex is the top-of-the-mind brand for people looking for a product to help them clean windows quickly.

Over the years, many marketers have attacked Windex's leadership position with products that claimed to work better: Glass Wax, Glass Works, and Glass Plus, among others. In my own experience, plain ammonia in water works just fine, and Glass Works (now renamed SOS Glass Cleaner), which contains vinegar, is a close second. But neither my opinion nor the blandishments of all the competitors have dislodged the name Windex and the reach-for-it reaction it triggers in shoppers' minds.

So valuable is a good trademark that it can be revived on new products even when the product it originally graced has gone by the boards. According to my former partner, Gene Whelan, when Carter-Wallace was looking for a trademark for a newly licensed antiperspirant discovery some years back, its own staff and its advertising agency came up with over one hundred candidates. The names deemed best were incorporated in test advertisements and checked for brand memory–retention with consumers. Included in the list was an old trademark of a product that had fallen by the wayside: Arrid.

When the research dust had settled, Arrid was the clear winner. It was not only a good suggestive name for an antiperspirant; it was a name that rang a memory bell for enough people to gain rapid acceptance. The brand became Carter-Wallace's cash cow in the toiletries field and remained a winner for over twenty years.

I have seen this pattern repeated consistently. If substantial advertising money is spent on a name, and if the original product under that name did not end in total public disgrace, reusing that name is usually better than launching an unknown new one.

Undoubtedly, the great value of old brand names lies in the public's persistence of recall for trade names and its fuzzy memory for specifics. In 1989 Canadian entrepreneur James H. Ting bought the Singer Sewing Machine Company for about $300 million. He not only got a dominant sewing machine business but a brand name to stamp on a range of consumer electronic products as well as on retail stores around the world, for the 140-year-old Singer name is established in more than one hundred countries.

Even the discovery of a little benzene in Perrier and the recall of that product or the terrible "Tylenol murders" and the subsequent Tylenol recall didn't shake the confidence of consumers in those two brand names. Both companies handled the problems forthrightly and the brand images survived essentially intact.

Not only do good trademarks help a company maintain profit margins and add substantially to the value of an acquisition, they can also be the source of unexpected revenue via licensing for noncompetitive merchandise.

Take the trademark Everlast, for example. Reminds you of boxing gloves, boxing trunks, and boxing headgear, doesn't it? But if you think about it, it does speak, flatly and clearly of strength and athleticism. And the mark, with its curved lettering topside, is distinctive. Why not put it on other sports clothing and equipment? Everlast Sports Manufacturing Co., rather than barge into product lines it wasn't really prepared to make and market, licensed the Everlast logo to U.S.A. Classic, Inc., and other marketers. At a 6 percent royalty rate, Everlast's licenses brought the company an annual six-figure income, most of which flowed directly to the bottom line.

Failure to seize a trademark opportunity can be the equivalent of conceding a product category. Potato chips were "invented," it is widely believed, in 1853 by an American Adirondack Indian, George Crum, who was the chef at Moon Lake House in Saratoga Springs, New York. When a restaurant patron—Commodore Cornelius Vanderbilt, no less—complained that his french fried potatoes were "not properly thin" and sent them back, Crum sliced the potatoes wafer-thin and deep-fried them. Instead of being offended, the railroad tycoon loved them, and the chips became a specialty of the Moon Lake House. They were known then as Saratoga Chips. Unfortunately, Crum didn't have the foresight or the business skill to seize that trade name and market the product.

Brand names in the potato chip field were decided by local entrepreneurship because early packaging was air-permeable, leading to stale chips and requiring frequent restocking of fresh chips. Thus, people in Dayton, Ohio, grew up thinking that "Mike-sell's" was the right name for potato chips; in California, it was "Laura Scudder"; in the New York area, "Wise."

It remained for Herman Lay to found a company in the 1920s to market the first widely sold packaged potato chips. While the brand name Lay's® Potato Chips still doesn't have quite the panache of Saratoga Chips, it's the top-of-mind name for most potato chip aficionados. In the $4-billion potato chip market that remains regionally fractionated, Lay's (now a division of Pepsico, Inc.) has about one third of the business, a leadership position it won by doing just about everything right, from standardizing its product to using freshness-protecting packaging to introducing eighty-five different flavors and shapes and colors, all tuned to local preferences and combined with a fabulous distribution system. No doubt, some of these varieties do not sell enough to meet large-company standards of efficiency, but cumulatively they are what gives Lay's Potato Chips its brand dominance, making it tough on competitors.

Trademarks are a deterrent to "me, too" competition but seldom a barrier to unique new products. In Chapter 9, you can read about how, by calling their product Smartfood, two guys from Hartford, Connecticut, turned popcorn into nutrition and themselves into millionaires. In the Trade Secrets section of this chapter, you can read how Stephen Bernard found a way of making potato chips taste so different that they sold themselves under the unlikely trademark of Cape Cod.

Getting the Best Trademark

If you wonder where in the world people come up with some of the strange trademarks you see every day, the answer is that it has become increasingly difficult to find a good simple name that is not already registered. So you see computer-generated words like UNISYS for computers, or deliberate efforts to be different like Peak Freans® for cookies, or made-up names like Haagen-Daz® for ice cream.

So important are good trademarks to savvy marketers that there are companies in the business of creating trade names. These companies get from $10,000 to $250,000 for name-creating and name-evaluating assignments. Other people generate long lists of name candidates using as sources their own people, their advertising agencies, and often their spouses and children.

Independent inventors and start-up companies, lacking such wide-ranging resources, characteristically settle for the name of the founder or whatever moniker gets attached to the invention during development as a product name. But remember, even though you have established a company under a certain name, that is no guarantee that your corporate name will be available to you for use as a product trademark.

The history of successful trademarks shows that no single selec tion approach is perfect. But perfect hindsight does disclose some basic characteristics of a successful mark:

- *It is simple,* which aids recall and makes it easier to design a distinctive trademark and package.
- *It is pronounceable.* People dislike names that puzzle them. It should both look good in print and sound clear when spoken. And if you expect to sell abroad, check both the pronunciation and the meaning of your brand in the languages of the countries where you may export or license.
- *It connotes good things for your product,* or at least doesn't turn off potential buyers.
- *It is available,* which means that many sources have to be checked: the U.S. Trademark Register, state trademark registers, trade directories and publications, association lists, telephone books, and so on. As a preliminary screen, you can check some of these sources yourself. *The Trademark Register* is useful in eliminating name candidates, but if a name isn't listed there, that doesn't prove that it's available. Or if you have access to Lexis or one of the other legal data banks, you can run a quick check of names through your own computer.

In the final analysis, however, retaining a trademark lawyer is a good investment if you expect your trademark to be a valuable property. For a serious inventor this means that, soon after the invention enters the product development stage, the trademark work should begin so that the product and the product concept, complete with name, can come together at one moment in time. Bear in mind that the 1989 change in U.S. trademark law enables you to file for a trademark before actually making a sale in interstate commerce.

Protecting Your Trademark

Unlike a patent, which is granted for a fixed number of years, a trademark is good for as long as you use it. But use it you must—use it by making and selling a product bearing that name in interstate commerce. When you do not, the mark can be considered abandoned and someone else can start using it. Furthermore, you must aggressively protect it against use by others, including the news media, which frequently use trade names without even starting the word or words with a capital letter.

Remember, too, that people can have regional rights to trademarks, a fact that underlines how important it is to check thoroughly before adopting a mark and spending money to promote it. Trademarks automatically come up for renewal; specifically, you must refile every ten years for any mark that was issued after November 19, 1989, and every twenty years for marks granted before.

Now all that is rudimentary. If you want a good elementary education on trademarks, read some of the books listed under that heading in the bibliography. *Copyrights, Patents & Trademarks* by Hoyt L. Barber is a good starting point.

However, the details of what constitutes proper use of your trademark, how to warn those who may use it improperly, and how to pursue those who may counterfeit your products and brand should be reviewed with your attorney. Just remember that your attorney can't help you unless you use that trademark actively and indicate that you consider it your brand. On the subject of using it, sometimes that's the only way you can get a trademark. Certain marks are not acceptable for the general trademark register, for example, those considered too descriptive of the product itself. However, even these names often become valid trademarks on the secondary register by virtue of enjoying unchallenged use for a number of years.

An interesting case in point is a product called Cot'n Wash®. This was a special formula of detergent chemically engineered for cotton sweaters and such. The name was obviously good for the product—too good, said the Patent and Trademark Office, too close to pure description. But since the product was already on the market, I advised the owners to continue using it and to feature the name while applying for a secondary register mark. It went by unopposed and eventually became a fully registered trademark, substantially increasing the value of the product to which it was attached.

It's up to the inventor/entrepreneur rather than the attorney to protect the trademark actively. Whenever and wherever you can, tell the world that you consider this trade name to be your trademark.

You do this by putting the letters *TM* (™) alongside your trade name before it is registered and the *R* in a circle (®) alongside it after your registration issues—on the product label, on the container it's shipped in, in your advertisements, everywhere. Without a registration, many lawyers advise putting the word "brand" alongside your trademark as well as the "TM."

And even that is not enough. Any time you see your trademark being used without a line acknowledging it as your property, you must protest promptly and in writing.

A landmark lawsuit that brought marketers up short on this subject occurred many years ago when The King-Seeley Company lost exclusive rights to the word *thermos* as the brand identifying its vessels that kept liquids hot or cold. The whole world had gone around saying, "Where's the thermos?" Newspapers used the word generically. And the court held that Stanley had not aggressively protested such use, thus dedicating the word to public usage.

Just as I have pointed out about patents, trademarks are a business tool, and companies use trademarks for different strategic purposes.

Du Pont is reported to have let the name *nylon* slip into the public domain because the company had strong patents that it was seeking to license. To increase the value of such licenses, Du Pont wanted the public to recognize nylon as a specific new fabric, no matter who produced it. It was a deliberate business decision.

Sony created a product category with the brand name Walkman®. The product itself was not technically an invention, but the product concept combined with this great name defined a highly portable tape player with earphones—thus providing Sony with a proprietary position but one that requires vigilant defense.

No company could be more sensitive to the problem of having its trademark become a common description for its class of products than the Xerox Corporation is. In an advertisement pleading with people never to use its name as a verb, "to xerox" in place of "to copy," the company pointed to the following now-common words that were once proprietary trademarks: escalator, cube steak, corn flakes, kerosene, nylon, yo-yo, linoleum, mimeograph, dry ice, high octane, trampoline, and shredded wheat. To which august list one might add aspirin, once the property of Bayer.

Don't let it happen to your trademark!

Copyright? Right!

That little *c* in a circle (©) followed by the year and the name of the copyright owner is so omnipresent in printed matter that it's easy to forget how useful it can be in protecting certain other products. Aside from its "normal" application to publications like this book, copyrights can provide certain kinds of protection for a variety of products.

Entertainment on videotape or film is a prime example. So are musical compositions on tape or CD or paintings and sculptures, all of which are, in the broad sense, inventions. For many such products, a copyright is the essential legal barrier against piracy—just note that "Warning-FBI" message on your next videotape movie rental.

For computer software, copyrights were the only hope for protecting programs (because almost all copy-protection inserts can be defeated) until the courts decided in the late 1980s that patents could be applied to software. In brief, a software patent is usually based on a system—the program combined with the hardware needed to run it—and thus cannot be evaded by small variations.

In contrast, a software copyright covers only the specific program as written; it protects against direct copies but can often be evaded by reasonable variations. Because computer software patents are relatively new, the precedents are still in the making, and the lawsuits are making the lawyers rich.

Every day there are headline events like these: Apple Computer sues Microsoft over its "menu" software, and the final outcome affects hundreds of inventors of new software programs who cannot afford such legal battles. Nintendo sues Atari and others over its copyrights on games for the Nintendo system. Nintendo also sues Lewis Galoob over a video device called Game Genie that makes it easy for mean little kids to beat the heck out of both of the Super Mario brothers.

And such court wars will grow like a computer virus because millions of dollars are at stake and it's largely new legal turf.

The costs of obtaining a copyright are quite nominal, and everyone who has an appropriate creative work should copyright it. But never forget that the copyright has value in only two cases: (1) where a weak competitor uses your property and then decides not to fight you, and (2) when you are prepared financially and emotionally to fight a major competitor.

There is a little-noticed use of copyrights that can, in the right circumstances, provide a useful picket for your fence. If you manufacture a tangible product—appliances, medical equipment, electronic gadgetry, for example—even when you are blessed with an issued patent, you can provide an added level of protection by copyrighting the instructions for use, assembly, or repair of the product. This is especially pertinent when you find someone from the Far East making a Chinese copy of your product and of the instructions, too!

Designing Design Patents

A design patent, mentioned briefly in Chapter 4, is issued by the Patent Office to protect against the copying of a product's *appearance* as opposed to its composition or structure or process of manufacture, aspects protected by utility patents. You can, of course, hold a design patent and a utility patent on the same invention and it often pays to do so, especially if the ornamental aspect of the product is distinctive. Incidentally, a design patent is good for only fourteen years versus seventeen years for other patents.

Because relatively small changes in design may avoid infringement, some people think that design patents are only good for framing and hanging on the wall and not for much else. That's a vast idea based on a half-vast understanding of the uses of design patents.

Basically, the strength of a design patent depends upon how important the appearance of the product is to its identity and to its strength in the marketplace. Does the particular look of this product help buyers to identify it? Is the look unique to the company's brand, thus reinforcing the trademark? Is the look relevant to the product's function?

Back in the 1950s, I was involved in a design patent situation that became the subject of litigation, a landmark example of how potent design patents can be as a business weapon. At the time, I ran an advertising agency, and Picturesque® stockings was my client. Picturesque nylons certainly were different. They were the first ladies' stockings with fancy designs woven into the heel and up the back of the leg. It gave men something to follow other than a plain seam. Retailers loved them; more important, they received a lot of publicity, women loved them, and business boomed.

In the soft goods world, where copying the latest designs seems like a birthright to most manufacturers, it was not long before other hosiery makers began to flood the market with similar products.

Stanton Sanson, the president of Picturesque and a most canny businessman, fully expected this. So, shortly after the original design was conceived, he had gone to his patent counsel, who advised him to file design patents on all the basic designs. When the imitators appeared, not only did Sanson have his lawyer notify those manufacturers that they were infringing his patents; he also sent letters to their retailers advising them that they should refrain from selling these infringing imitations lest they too be subject to legal action. He also ran full-page advertisements in the trade press explaining his proprietary position to the entire stocking world.

Most manufacturers backed off. Raven Mills in North Carolina, however, either didn't believe him or had counsel who needed the work. Stanton Sanson sued in North Carolina, where one might be concerned, not unreasonably, about local bias on the bench. But Sanson had a trump trick: He asked me to prepare a motion picture film showing women, just from the knees down, walking away from the camera. Some of these women wore Picturesque stockings with the patented designs; others were wearing the defendant's stockings.

When that film was shown in court, with the design patterns in motion rather than just draped over a courtroom table, the design infringement was dramatized. The real-life use of the product was replicated, and the close imitation became an infringement in the eyes of the court. How was it possible to get permission to show that film in court? That's what smart lawyers are paid to accomplish.

The Picturesque case is now in the patent files of the National Archives.

That example should not be taken to suggest that all design patents can be successfully enforced. To begin with, the Patent Office just doesn't question design applications as it does utility patents. In many cases, design patents cover items that are very easy to design another way with no loss of either usefulness or beauty. Take a 1990 patent, No. D306,811, issued to Richard DeLeon. This covers a bathroom wall fixture that will hold toothbrushes in conventional small slots on the sides and a toothpaste "pump" container in a largish receptacle in the center.

From a marketing standpoint, even granting that some people may wish to get the toothpaste pump off the basin top, it's a considerable challenge to persuade people to replace their existing toothbrush holder or to add an extra one. But even if the product succeeded, my eight-year-old granddaughter, by changing any of half a dozen shapes or dimensions, could design a perfectly good holder that would be clearly distinguishable from Mr. DeLeon's. Small wonder that some design patents are of little help in court.

Furthermore, even with a design patent that covers a genuinely unique form and appearance, one must face the task of enforcing it. You've seen the Itty Bitty Book Light, the articulated reading light that comes in a package designed like a book. It's the product of Zelco Industries, Inc., a company run by Noel and Adele Zeller, who, with the help of some advanced bulb manufacturing from Japan and good industrial designers in the United States, created and launched the product in 1982.

The product was a huge success. The designs of both the light and the "book" package were so beautifully simple and pure that the

Museum of Modern Art included the product in its permanent collection. And the Zellers sought and obtained a design patent.

Less than twelve months after its introduction, imitations from the Far East flooded the market. Zelco sued. Some cases were settled out of court for royalty payments—but not for damages for business lost on a hot item while waiting two years for court action.

Noel Zeller also had a mechanical patent and copyrights and trademarks. But he had an unusual problem. He had a too-successful product. As he explained to me,

> At one point there were more than seventy-two imitators around the world. We sued companies not only in the United States but in Australia, Canada, Germany, France, etc. We spent over a million dollars in legal fees. Even if you got a judgment, the small companies just went bankrupt and the big ones appealed. My wife and I realized that the only true defense is to be very effective with the marketing of your product.

Despite all that, between legal actions and aggressive moves in the marketplace, Zelco has retained perhaps 80 percent of the market for tiny reading lights. And it is quite profitable.

Significantly, the Federal Appeals Court in Washington has upheld injunctions based upon design patents where it was clear that the patent holder might face immediate economic loss unless the alleged infringer was stopped. For instance, the Avia division of Reebok International, Ltd., used a design patent on its running shoes to bar L. A. Gear from marketing close copies. Such injunctions, while perhaps not offering the pure satisfaction of a judgment for damages, do cost the infringing party heavily in the marketplace and therefore constitute a valuable tool in any business where time is critical.

I am surprised that many manufacturers apparently fail to pay attention to protecting small but valuable differences through design patents. For the sake of discussion, have you ever noticed the design of the rear side window of the BMW? At the bottom of the trailing edge of the window, the chrome frame, instead of forming a sharp corner, bends obliquely, creating a graceful "open" motif. This is a nuance of design, but it was unique when introduced and it works perfectly with the minimalist styling of the rest of the car.

Now the next time you see a 1991 or earlier Acura Accord or Legend, check that rear window shape and the chrome trim, and the shape of the rest of the car for that matter. Maybe BMW doesn't want to hassle Honda or maybe it doesn't believe in design patents or maybe its lawyers are too busy. But wouldn't it have been fascinating if BMW had taken side-view movies of the BMW 525i and the Acura Legend—the technique we used with Picturesque stockings—and asked a court for an injunction to stop Honda from marketing a "confusingly similar" vehicle?

Of course, the last two paragraphs are simply speculation.

But not to worry. BMW will survive this apparent oversight. Nonetheless, if you're an inventor/entrepreneur or a corporation president-to-be, anytime you're involved with a product where styling is a critical selling element, give some thought to designing with a design patent in mind. It's an underutilized protective picket.

Trade Secrets—Sssshhh!

This is an area that can indeed be a slippery slope—because everything depends on the specifics of a situation.

Whereas patents rest on statutory law, trade secrets are based on the common law principle that you cannot steal the tools a person uses to earn a living.

Even though the laws concerning trade secrets vary from place to place, you can start with this general definition in the federal courts and in most states: "A trade secret is any formula, pattern, device or compilation of information which is used in one's business, and gives him an opportunity to obtain an advantage over competitors who do not know or use it."

Another way to identify a trade secret is by its genre:

- *Secret technology*. Broadly, this covers any piece of information that is not known in a certain trade or industry and that provides a competitive edge. Specifically, it can be a formula, a manufacturing device or method, a way of treating or sorting or packaging or transporting materials—and the list goes on and on.
- *Secret business knowledge*. This is even broader and probably harder to codify and protect than technology. It includes such items as customer lists, market research, advertising response, sales routing—many items that directly affect the marketing of future inventions.

Many famous trade secrets relate to the food industry. Coca-Cola's formula is considered highly proprietary by that company and is guarded by a series of codes for the ingredients and secrecy agreements with employers and suppliers. I was once told by an officer of Hershey that Hershey's chocolate is unique because of the enzymes in the old mixing vats, which for many years were zealously maintained in their original condition.

A "secret process" need not be secret forever in order to serve the business purposes of a company. When Stephen Bernard began making Cape Cod Potato Chips, he took a giant step back in time, using a batch-fry method that had been abandoned fifty to sixty years before as too slow and too clumsy. Then he selected cottonseed oil to cook with because it contains no cholesterol and permitted him to avoid washing the chips after the frying process.

As Bernard described it, "It's a slower cooking process done at different temperatures. We get a hardier and tastier product. Also, because we don't send our chips through a water bath as others do, they retain a lot more flavor."

With a small business on Cape Cod and its vicinity employing a very small local work force, it was possible to keep the exact manufacturing process quiet. But as Cape Cod Potato Chips caught on and it became necessary to build a large production plant, Bernard faced two dilemmas: All high-speed chip equipment was designed for continuous-flow production, and he could not get the Cape Cod taste without the batch process. He solved the problem and created another "mystery factory" by going to a variety of people who make equipment for the potato chip industry and ordering one item here, another there—all designed to automate the plant completely from the moment the potatoes are dumped from a truck until they are packaged and in cartons, *except* for that small-batch cooking right in the middle of the line. There he added automated weight hoppers and computerized belts and extra kettles to match the capacity of the rest of the operation.

In effect, Steve Bernard created a new way to make potato chips while using an old manufacturing method. He created a bundle of manufacturing know-how, not patentable, but very hard for another chip maker to duplicate without a copy of the plant engineering plans. In due course, Cape Cod Potato Chips were copied by small makers with small kettles and long memories, but no major competitor surfaced.

In 1985, five years after its inception, Bernard and his wife sold their company, then producing some 3 million bags of chips a year, to Eagle Foods, Inc., for a kettle of cash and stock options. Eagle has spread the distribution to 85 percent of the United States and parts of Canada. The Bernards have gone on to start a new business making croutons and bread crisps under the brand name Chatham Village. But had they not hidden so quietly the way they made an old chip do new tricks, the Bernards might still be selling potato chips from a retail store on old Cape Cod.

There's a basic rule you should remember about trade secrets. When such cases get to court, the decision will almost always depend on your prior efforts in identifying the information as secret and your taking appropriate action to protect it.

From an employer's standpoint, it means not only appropriate confidentiality agreements with employees, but providing a secure place for secret developments and a record of everyone who sees the developments—and seeing to it that the information is locked up at night. In this era of industrial espionage, it often means having an internal security force to mind the store. Stealing secrets is not confined to military weapons, as witness the theft of precious formulations for new drugs from such stalwarts as Merck & Co. Unless the employer creates a record of due diligence in identifying and protecting trade secrets, the employee cannot be expected to sort out what's proprietary and what's not.

Then, of course, there is the disaffected employee who turns into a thief. That happened to the 3M Company when a contract employee, Philip A. Stegora, got his hands on a revolutionary tape developed for doctors' use in putting casts on broken bones. Stegora mailed samples to four of 3M's leading competitors with a demand for $20,000 as the price for details on the technology. None of the four informed the 3M Company about this; but at Johnson & Johnson, a product manager sent the samples to R&D, where they were tested and analyzed—and knocked off. 3M sued and was awarded $100 million-plus for loss of its trade secret.

As attorney Stan Lieberstein points out in his book *Who Owns What Is in Your Head?** trade secrets are lost when the subject is ignored until after the apocalypse:

> C. Lester Hogan resigned as executive vice-president of Motorola to become president and chief executive of Fairchild Camera and Instrument, and brought along with him seven top men from Motorola. The impact of this personnel shift was such that within a week Motorola stock dropped in value by 10 percent. Motorola's response was to sue. The company sought an injunction prohibiting disclosure to Fairchild of any Motorola trade secrets. Motorola eventually lost. Largely because of the way the information was

*New York: Hawthorn Books, 1979. Regrettably, out of print but still to be found in some business book libraries.

> handled internally, Motorola was unable to prove that the information its former employees could pass on *was* a "trade secret."

From an employee's standpoint, the question of "what's mine and what's theirs?" is one that every prolific inventor must consider, preferably before signing an employment agreement.

Ever since the landmark *Norfolk v. Peabody* case in 1868, the courts have recognized that you can't walk out with your employer's trade secrets to start your own business or to work for a competitor. On the other hand, an employer can't stop a person from moving to a new company and using either the skills learned during his or her employment or ideas generated outside the job assignment. In practice, it's the specifics, not the principle, that decide cases. When there could be reasonable doubt as to who owns an invention created during employment, the right move, the only practical move, is to get an agreement *before* doing anything that would make the discovery more valuable and worth fighting over.

A great example of the wise application of this rule was set by Harvey Phipard, the man who invented the multimillion-dollar sheet metal screw while employed in a plant producing such products.

Harvey described his experience this way:

> When I was graduated from MIT, I could have worked for Bethlehem Steel or Westinghouse or Singer Manufacturing, but they all required that I sign a statement that anything I invented belonged to the company. I didn't think I was going to invent anything in particular but I thought that was unfair. I would be giving away something and I didn't even know what.
>
> Then Continental Screw told me, "Sure, anything you invent belongs to you until we make an agreement on it." So I went along, and I invented four or five little things that were unimportant, but they paid me $100 or $200 a year and things went along fine—at least until I invented the Phipard screw."

How Harvey Phipard made millions under his agreement-to-agree is told in detail in Chapter 10.

The principle is simple: What are you being hired to do? What is within the scope of your employment? Your answer to these questions requires forethought and needs to be reflected in your employment records.

More difficult still are situations where the "inventor" is an independent contractor, as is frequently the case in the development of computer software projects; both parties have to be concerned about who owns any "intellectual property" that may result from the work. So it's essential to have an advance agreement defining what is being bought and paid for and what happens to any collateral discoveries that may result from the work.

Looked at from the viewpoint of the inventor or a start-up company seeking to license a new product, trade secrets serve a particularly valuable function. They not only reinforce your patent but give strength to your negotiating position. And, as detailed in Chapter 11, granting the right to use the know-how separately from the patent grant can preserve your royalty, at least in part, if the patent rights fail.

Inventors and businessmen alike are paying more attention to intellectual property as assets to be protected. You could get a good start on the subject with the Lieberstein book mentioned earlier and a broad education with *Protecting Proprietary Business Information and Trade Secrets* by James H. A. Pooley.

Buying Instant Exclusivity

Famous names, attached to new products, can buy not only rapid buyer recognition but also a certain cachet and exclusivity.

Endorsement by famous athletes, models, rock stars, and actors can ensure a certain amount of attention for a product, perhaps even some additional distribution and sales. However, this has its risks. People, particularly high-profile people, can lose their popularity and drag their endorsed products down with them. When Bo Jackson's hip problems knocked him out of football and baseball simultaneously, he lost some value as a spokesman for athletic shoes. When Pete Sampras was unable to follow up his U.S. Open tennis victory with any other for a year, Wilson and others who had rushed to sign him could only pray for more net cord shots.

However, tying into famous names who can bring authority to the product—when it works just right—can indeed be a shortcut to consumer favor. Elsewhere, I mentioned how Chef's Choice knife sharpeners was able to get an endorsement from Craig Claiborne, longtime *New York Times* food columnist.

Calling a perfume creation Elizabeth Taylor's Passion endowed a chemist's olfactory compound with a certain assurance of sexiness beyond what the ordinary sense of smell might discern. And this coupling of fragrance and femme fatale proved extraordinarily profitable for Elizabeth Arden.

Again in the beauty business, take the case of Gerry Rubin, who wanted to move his business from supplying beauty salons with hair dryers and hair-styling tools to mass marketing these products to consumers. Rubin knew that he had a quality hair dryer, which would translate to higher price, not exactly a blessing in a category where cheaper and cheesier had been the rule. Rubin and his partner, Aaron Shenkman, licensed the Vidal Sassoon name for hair dryers, outbidding Clairol and Gillette by offering $100,000 at signing and 10 percent of sales versus a 5–6 percent normal royalty. As Rubin saw it, quite correctly, the royalty was the cheapest advertising he could buy, and the quality product plus the Sassoon name would command a higher price.

As luck would have it, Vidal Sassoon sold his line of hair shampoos and conditioners to Richardson-Vicks, Inc., and then RVI was taken over in 1985 by Procter & Gamble. P&G promptly began spending $30 million a year advertising Vidal Sassoon shampoos and conditioners—and isn't that a fine form of brand building for a line of hair dryers?

On the rare occasions when famous people literally create products, their names provide a protection against copying as real as a design patent. Such a happy circumstance occurred when Nan Swid and Addie Powell convinced twelve of America's most famous architects to design a line of cups and saucers and plates for their start-up tableware company. Such renowned architects as Philip Johnson, Richard Meier, and Stanley Tigerman—all designers of world-famous buildings—signed on to create Swid-Powell's new line of china. The combination of good design and great provenance swept the company along for an annual sales volume of over $5 million. Then came the knockoffs of their designs from the Far East. They found themselves without patents or any other strong legal defense. Wisely, they kept their stable of architects and turned to custom designs for upscale housewares and linens.

One last example: You need not necessarily pay for a famous name if you are smart enough to create one or appropriate one that everybody *thinks* they have heard before. A case in point is The Franklin Mint of Philadelphia, a company marketing "limited edition" pieces of special-interest art. Franklin? Yup, we've heard of him. Inventive fellow, wasn't he? Mint? Who could be against the Mint, and isn't that located in Philadelphia, anyway? Must be reliable people. And enormously successful, too.

Other Pickets for Your Fence

Military men are likely to talk of "lines of defense." It's a fair metaphor for businesspeople to follow. A variety of proprietary positions builds a combined strength that's greater than the sum of the parts.

This thinking is especially applicable to inventions and new-new products because, like all young things, they require development time. Thus, you should never rely solely upon a patent or a trademark or a copyright or famous personalities if you can generate a combination of these set-aparts. And you can add other less obvious but still valuable forms of protection. Here are some supplementary, synergistic lines of defense:

- *Governmental approvals.* In an age when more and more things are subject to local, state, and federal regulation, owning an approval or permit can buy lead time. Some companies even work with government officials to write the regulations.
- "*Black boxes.*" A black box is simply an enclosure that no one but authorized personnel can look inside. Black boxes can be used two ways.

I've been in a factory where there was a completely enclosed room in the middle of the plant floor. There was only one door into that room and the sign read "Positively No Admission Without Personal Code." The production line went into an opening on one side of the room and came out from an opening on the other. But the electronic units being assembled took much longer to emerge from that other side than direct travel on the moving line would have suggested. And when the product came out, it was covered with a black metal enclosure. To this day, I am not sure what—if anything—was going on in that room. But I do know that the visitors from around the world who went through that factory came away deeply impressed. And the product was never knocked off; it died a natural death of obsolescence.

It may pay in licensing a product to present your invention via a "black box" technique—especially when revealing the actual mechanism of action would be too dangerous for a first meeting, even under a confidentiality agreement. You need not always resort to a physical enclosure to use the black box principle. Once I demonstrated a fast-drying hand cream by having a young lady use it and then, thirty seconds later, run her hand down the suit lapel of the president of the company we were presenting to.

• *Trade agreements.* Taking the precaution of checking with your favorite lawyer, you can, under the right circumstances, block or slow down competitors by making exclusive or preferential agreements with suppliers of raw materials, contract manufacturers, and major distributors. The Interplak toothbrush works as it does because the developers found that there were only two places in the world where it could be made and they took the entire production of the better one. When Dynagran Corporation was developing a calcium-supplement product for Marion Laboratories, the cost of materials was a serious consideration because Marion would be paying a royalty. My partner in that company, Bill Valentine, found a source in Vermont where commercial-grade calcium could be converted to pharmaceutical-grade at a very favorable price—and we kept that mine under wraps for quite a period of time.

• *Public relations.* There are things that good press stories can do for a product that will cling to the brand image and last for a long, long time. Public credibility, built up by praise, even faint praise, from reporters, is often a better defense against imitation than paid advertising. When Dick Finnis, Paul Ross, and Fred Kulow set out to introduce Bee Pollen of England in this country in the 1970s, they had a product with an odd name, unkown in the United States and always in danger of running afoul of the FDA on claims. They hired Les Lieber, an old pro in publicity, to bruit the name Bee Pollen around. Lieber found enough athletes who yearned for an extra "oomph!" to try the product. Among them was Steve Riddick, one of the world's "fastest humans," and Bee Pollen's reputation as an energizer was made.

When you garner sufficient publicity for a product, it is a giant boost for a trademark and helps to preempt a field. No better example was seen in the late 1980s than Rollerblade®, the in-line roller skate that pioneered a category worth an estimated $150 million in 1990, and held two-thirds of the market, primarily through demonstrations, staged events, and press coverage.

Think about product protection this way: If you have both patents and other proprietary picket fences, it becomes geometrically harder for the would-be imitator to say to his engineers, "Design around that!" or to his lawyers, "Break that patent!"

All the protection fences combined, of course, are not a substitute for genuine product breakthroughs, innovations that earn their place in the market. But when the product performance and the product trademark and the product design all merge elegantly, *and* you have a protected position, there's a synergism that's hard to beat. Consider the staying power of products like the Mont Blanc Diplomat Pen, the Cartier Santos Watch, Hershey's Kisses, or the old Zippo Lighter—or Tylenol or Polaroid or Xerox.

Chapter 6

Capitalizing on Intellectual Property

The words *intellectual property* are used by lawyers and other reasonably educated people as a catchall phrase for almost anything that isn't tangible, such as real estate, machinery, and inventory, for instance. Intellectual property includes writings, notes, drawings, sketches and designs, trademarks, computer software and know-how, as well as inventions, whether mechanical, chemical, or biological.

Thus far, we have walked—or perhaps run—through some of the fundamentals you need to know to refine a discovery into nuggets of gold.

Now let's take that giant step into the real world of making money from inventions, a world where the winners invent solutions to problems relating not only to patents, trademarks, production, and marketing but also to the natural human inertia that so often slows new developments. I want to try to give you a gut feeling for what it takes to turn inventions into millions. For the creator or the promoter of an invention, whether within a corporate environment or working as an individual, the number one priority must be to get the invention out of the intellectual stage and into the income-producing property stage... *without losing proprietorship to outright thieves or via a clever legal proceeding.*

So raise your right hand (hold the book in your left) and embrace the inventor/entrepreneur's credo:

• I shall find a way to make this [choose one] invention/technology/product as important to other individuals or groups as it is to me. I shall, judiciously, give a piece of the action to anyone who can help move my property to market; and I shall beg, threaten, exaggerate, cajole, or otherwise snare anyone I need for my team. I cannot do it alone and do not intend to.

- I shall move, react, change to meet situations as they arise. My only fixed policy is to survive.
- I shall defend my proprietary rights aggressively against intruders, using the best lawyers who, for whatever reason, consider my cause terribly important to their reputations or pocketbooks.

Of course, credos must be translated into judgments and actions. That's what the rest of this book is all about.

In this chapter, you'll read three in-depth success stories, each quite extraordinary, each chosen to give you multiple insights into the practicalities of turning inventions into income.

It took Gordon Gould thirty years to gain control over his laser patents, which will bring him $60 million. Harvey Phipard turned his invention of a "simple sheet metal screw" into an annual income of $1 million, even though he worked for a company that made such products. And Jon Lindseth and his team needed only four years to turn George Clemens's electric toothbrush invention into a company that was acquired by Bausch & Lomb for $133 million.

Patent, Patent, Who Gets the Patent?

The prime question on every invention is who really invented it. This is closely followed by who is going to get the patent and what will happen if the patent, when issued, is challenged by an interference action, usually filed by someone claiming priority of invention, or by a reexamination action challenging the judgment of the patent examiner.

Under United States law, the patent goes to the person who made the discovery first, not the person who filed his or her patent application first.

As every good scientist knows, this dictates the keeping of meticulous records as to the steps taken to arrive at the invention and the precise date, even the time of day, of the invention. And such detailed records must be kept in a bound notebook, not a loose-leaf, be written in ink that is not erasable, and have more detail rather than less.

No one is more sensitive to the importance of recording discoveries than Gordon Gould, whose patents dominate the field of lasers, patents that became truly his only after a thirty-year war with the Patent Office and a consortium of corporate opponents, patents that have already brought him $2.3 million before the first license was ever signed, plus royalties that should, he estimates, amount to some $60 million in his lifetime.

When Gould was a graduate student at Columbia University in New York, he spent much of his time trying to build a working laser, despite the academic objections of his mentor, Dr. Polykarp Kusch. It was a rich creative period at Columbia. Another professor, Dr. Charles Townes, and a fellow doctoral student, Arthur Schawlow, also worked intensely on lasers. These two were destined to become Nobel Prize winners and, since they married sisters, brothers-in-law as well.

At Columbia, Townes, I am told, held regular Friday afternoon teas for students, who would come and talk about their work. Gould attended, often. He held forth on his preoccupation: applied science in the field of lasers.

One night in that era, November 9, 1957, Gould, awake in the wee hours, suddenly saw the entire laser process clearly. He immediately began writing notes and making sketches, describing both the invention and its future uses. He continued for several days, visualizing various embodiments of his laser. Then he took the notebook to a nearby notary public, who dated it and stamped it with his seal.

As the laser maze unfolded in the years that followed, Gould faced a convoluted fight with the Patent Office and five different patent interference actions instigated by competing inventors and angry major corporations. At a critical juncture, in an action versus Control Laser Corporation, the question of invention priority was at issue. The chief witness for Control Laser was Charles Townes. As the cards were laid on the table, a pattern emerged: Gordon Gould's own notes, point by point, preceded Charles Townes's by approximately thirty days. References in Townes's own notes, such as "Gould thinks that thallium will work," strongly reinforced Gould's claim.

Gould's long legal battles were caused in part by his failure to follow up his discoveries with timely patent filings. It was, Gould told me, a communications error. He had retained a young patent attorney, Robert Keegan, who told him that his invention must be "reduced to practice." Gould took this to mean that he must have a working model, and there were many obstacles, such as money and time, between him and a functioning prototype. So he delayed filing an application.

The world, of course, keeps turning. In 1960, Townes and Schawlow received a patent on their optical maser (masers amplify microwave energy, whereas lasers amplify light)—but they were widely regarded as the fathers of the laser as well until Gould's final legal victory some twenty-five years later.

And here is an interesting sidelight: Although Townes and Schawlow were also working on laser inventions, it was neither they nor Gould who built the first working prototype. That honor went to Ted Maimon, a research scientist at Hughes Aircraft who built a ruby laser despite management's orders to kill the project. (Apparently, Hughes had a penchant for killing laser projects in those days; Mary Spaeth, inventor of the tunable dye laser, says she completed her invention at Hughes in the face of a direct order not to work on it.) Ted Maimon, so the story goes, from a wide choice of rubies picked by instinct or by luck precisely the right one for a laser. Then, when his prototype still didn't work after days of trying, he banged the lab bench and the laser fired away. Whatever the elements of science or serendipity in Ted Maimon's work, he got better results than some very talented people at Bell Labs, Westinghouse, and other major companies who were simultaneously seeking the secret of a working laser.

But Gould's discovery of the actual operating principles of the laser clearly preceded Maimon's model. And he eventually learned what none of us should forget, that "reduction to practice," as used in patent law, simply means describing an operable version of the invention, not necessarily actually doing it. It is a reality check on the claims intended to help the patent examiner.

Ultimately, Gould's failure to be first to file—and other legal ins and outs—cost him important arenas of the laser world; he won three and lost two of the interference actions brought within the Patent Office.

He received priority over a scientist at Westinghouse who was doing laser development work. He lost in significant areas to Townes: Portions of Gould's application were struck off, including all low-power uses of lasers, such as metrology and general measurement and communications uses. However, he retained all high-power uses because neither Townes nor Schawlow had foreseen in their applications that a laser could develop such power. And in terms of the most important commercial applications, it turned out that gas-operated lasers, used in optical pumping for rubies, and other high-power lasers were the first to achieve big commercial success.

During Gordon Gould's long walkabout through the patent desert, he had many challenges, many opportunities to lose it all. The misunderstanding about "reduction to practice" was only one. As recounted in the pages that follow, he almost lost the laser rights to an employer and he was blocked by the United States government at a time when the country desperately needed leading-edge technology.

Soon after he left Columbia, Gould went to work for the Technical Research Group, Inc., which became a division of Control Data

Corporation. Under TRG's auspices, he applied for a government grant for $1 million to develop his laser invention. Fortunately, it was approved. Unfortunately, while reviewing the application, the government minions discovered that, some years earlier, Gould had had a left-leaning girlfriend with whom he had attended a course on Marxism and whom he later married. That was enough for the government to blacklist him. Gould was barred from working to develop applications of his own invention. He lost more years in the patent fight, and the government lost a prime opportunity for ascendancy in advanced weapons systems.

Gould recovered his laser patent applications from TRG, only to face years of Patent Office rejection. At a point where he was almost ready to abandon the chase, the wheel of fortune turned his way. In 1976, Gene Lang of Refac Corporation sent him to Joe Littenberg, a senior patent attorney at Lerner, David, Littenberg & Samuel. Buried in the multiple claims in Gould's patent filings were two claims for creating lasers, one by optical amplification and the other by gas discharge. Littenberg spotted these and recognized that these methods, and not just reflecting mirrors, were basic to high powered lasers. Within a year, the first Gould patent issued, setting the laser world aghast. But it was not until 1988 that the last of the patents was won.

In the interim, many actions were argued with the Patent Office and in court. In 1986, the Lerner, David firm went to federal court to force the Patent Office's hand on the gas discharge laser patent. They lost and filed for reconsideration. The response from the Patent Office was practically an invitation for third parties to file for reexamination of all of Gould's patents. An avalanche of such requests followed. Among the filers was General Motors, then a licensee, which had insisted on putting a clause in its license permitting it to challenge the patent.

Although patent files normally go back to the original examiner for a second look, Gould's applications were assigned to Harvey Behrend, an old-timer with a tough-guy reputation. Behrend sat on the matter and then, in 1983, rejected two of Gould's already issued patents. Soon, the Patent Office decided it would consult with the Justice Department because the Gould patents might be enforced against the government itself. (Note: the government is not exempt from paying royalties on U.S. patents.)

Then in December of 1985, the tide suddenly turned Gould's way for good. Judge Flannery of the district court, who had previously ruled against Gould, now issued a strong favorable decision. He ordered the Patent Office to issue the gas-discharge laser patent.

Then the Patent Office's appellate board reversed Behrend's decision in the reexaminations of the optically pumped laser and the "use" patents. Other patents that had been buried in examiners' stacks surfaced and were issued.

At long last, in 1986 the suit against Control Laser Corporation, the Florida laser manufacturer that had been the stalking horse for the entire laser industry, went to trial. Legally, the issue turned on the validity of the optically pumped laser patent. That was when, as mentioned earlier, Charles Townes, Control Laser's star witness, was faced with a comparison of his notebooks and Gould's. And that, combined with Gould's victories in the reexamination process, deter mined the court's decision. Having in effect bet the company on this protracted litigation, Control Laser shortly thereafter went out of business.

Even knowing that one is right, how does anyone maintain the fortitude to pursue the grail for almost three decades? Gordon Gould, right from the start, found friends who supported his cause. Although he could never be sure how it would turn out, Gould firmly believed he was doing the right thing. When I asked him, "Did you ever say to yourself, 'Gee, if I had known then what I know now, I would have done things differently?'" he replied, "If I had had any idea of how long it was going to take and the difficulties that would arise, I'd have dropped it long ago. But I had confidence that, in the end, I *would* get these patents and that it was always just one or two years away. So it was in all innocence that I kept on with it and I shouldn't be credited with all this bulldog determination that carried me on for close to thirty years."

Gould's Lesson Legacy

Gordon Gould would not have "stayed alive" to see any of this happen were it not for his carefully detailed and notarized original notes—a lighthouse lesson for all who would make millions from inventions.

Nor would he have made it without enlisting the help of others and sharing the opportunity with them.

His ultimate victory was made possible by two fundamentals that every inventor/entrepreneur should take to heart:

1. *Write it down.* In words and pictures. In much detail. Then don't lose it. Get it notarized or give a copy to your patent lawyer. Gould was forced by the government to surrender his notebooks when

he was working for TRG in the Red witch-hunt era. But he kept a copy anyway, which saved the day as to the date of the invention.

2. *Make friends, find strong allies, and especially make sure that your patent attorney is dedicated to your cause.*

How Gould followed these principles is worth some discussion. Gould found *two* patent attorneys who made all the difference for him. The first was Robert Keegan. Gould told me,

> He actually came out of law school and started working on this. The first half of his career was all wrapped up with these patents. Now, he was not the most capable person, not brilliant, but he kept on with it. Even after he left his law firm, he undertook for several years, without pay, the struggle to keep up the patent prosecution—before I met Dick Samuel. I promised Bob 10 percent if we ever got anything out of it. Recently, we bought him out for a half million dollars, and it was my sweet pleasure.

(Oddly enough, this was the same person who let Gould labor under a misapprehension as to the meaning of "reduction to practice.")

The second "attorney" was the group at Lerner, David, Littenberg & Samuel. Sid David, William Metlik, and Roy Wepner, as well as Joe Littenberg and Dick Samuel, all worked actively for Gould, both in arguing with the Patent Office and in the innumerable court cases. Talk about showing faith in the future!

A nice sidelight on how things happen was Gene Lang's comment to me when I asked him why he had agreed to back a dicey patent chase like this one: "I just couldn't resist the idea of owning a basic piece of the light spectrum."

The law firm was tremendously impressed by Gould's thoroughness, from the original notebooks right on through to his unceasing efforts with the Patent Office and undertook to handle the litigation on a contingency basis, getting 15 percent of the take from Gould and another 10 percent from Lang.

Eventually, Samuel left his law firm and set up Patlex Corporation, a vehicle conceived to attract investment and to finance the continuing costs of litigation. Between the original legal services that Gould bought with a percent of his future and the money Lang put up, Gordon Gould, who certainly did not have the wherewithal himself, financed the first phase of the legal battle.

But more money, much more, was required. So Dick Samuel

conceived the idea of setting up a public corporation to own the patents and sell stock to the public. As he cast about for a vehicle, he met Gary Erlbaum, owner of a small Philadelphia public company named Panelrama Corp., a chain of building supply stores. Samuel explained the potential of lasers and Erlbaum was enchanted. By happy coincidence, Erlbaum was becoming disenchanted with the building supply business. They struck a deal. Erlbaum sold off the stores, keeping only the real estate the company owned. Samuel, his partners, and Refac, together controlling 64 percent of the laser patents, put that asset into Panelrama, now effectively a corporate shell, getting 64 percent of the stock of Panelrama, which then had over $1 million in cash in the till.

Shortly thereafter, the name of the corporation was changed from Panelrama to the Patlex Corporation. The timing was exquisite. In 1979, after Gould's second laser patent issued, Patlex sold the public "shares in the future of light" and raised $6–$7 million. Gould was able to cash in. He sold Patlex 15 percent of the royalty stream for $2.3 million, leaving himself 21 percent that he could tuck away in the safe. That gave Patlex a war chest with which to fight one of the great legal battles in patent history.

As a practical matter, however, in 1985, when the basic "method" patent was issued, Patlex knew it couldn't license laser manufacturers—because the patent reexaminations were still going on. So the Patlex people devised a brilliant strategy: Sign up laser *users*.

As Steve Nagler, who was then house counsel for Patlex, explained it to me:

> We went to GE, to Motorola, to Ford, and others and offered a license based upon 5 percent of what they had paid for lasers they were using, retroactive to the date of issuance of the first patent in 1977. It was a companywide license. Most of those licenses also said that if they bought any future laser from an unlicensed manufacturer, they would pay a 6 percent royalty.
>
> It took a lot of negotiation. For instance, when we walked into Ford, we had no idea of how many lasers they had on their lines. We had scraps of information from industry magazines, so we made rough estimates and relied heavily on the good faith of our counterparts in the negotiations.
>
> Under the circumstances, we got a very fair shake. That money was the difference between being able to continue or not with our litigation against Control Laser and

> the gas-discharge laser patent case, which was then in court, and the reexaminations that were still going on in the Patent Office.

Altogether, the use licenses were not small potatoes. The Ford agreement alone was worth over a million dollars and the total reached over $9 million. Patlex was on a roll. The first manufacturing license was signed with United Technologies, which, if not a big laser maker, was certainly a prestige name. By December 1985, when Judge Flannery's decision came down not only sustaining Gould's position but excoriating the Patent Office for its obfuscation and errors, Patlex could at last turn its attention to making the most of the cash that would assuredly flow in.

The Guys in the White Hats

Never forget, every inventor needs a champion, not just a lawyer. It was Dick Samuel's unrelenting assault upon the tactics of the Patent Office that ultimately won the day.

Gould's own credibility attracted other champions. Keegan, Lang, Erlbaum, and others less prominent but still important, are shining examples of what even a great inventor needs: True believers who express their belief through action.

One more person stands high in Gordon Gould's esteem. When Gould was working at TRG, the company was owned by Control Data Corporation, and Control Data decided to sell it off, lock, stock, laser patents, and all. Such was Gordon Gould's good fortune, however, that the president of TRG, Dr. Lawrence Goldmuntz, had recognized when he hired him that Gould was bringing his prior work on lasers to TRG and let Gould retain certain rights in the patents. This effectively precluded Control Data from selling off TRG while Gould's patent rights were on its list of assets. A deal was struck and Gould bought all the rights to his patents from Control Data for a modest amount of money.

"It was a very upright and courageous thing to do," Gould told me with feeling. "I was so innocent, I could have lost everything. It was Doctor Goldmuntz who ensured that, and I have a warm spot in my heart for him. I mean, most businessmen simply aren't that ethical."

Of course, what goes around comes around. From Gordon Gould's first laser flash of insight in 1957 until Judge Thomas A. Flannery's favorable decision in 1985, scientists had been inventing practical applications for lasers. In surgery, in manufacturing, in musical reproduction, and on and on, equipment using lasers became the cutting-edge technology for product breakthroughs. Thus, by the time Gould and Patlex finally acquired clear patent rights, there were plenty of candidates for licenses. Had Gould's patents been granted, say, in 1960, the prescribed seventeen years of patent life would have passed with few laser products on the market and sparse royalties for Gould.

And now as Gordon Gould sits back in his home on Virginia's Eastern Shore, you can just tell he's enjoying something that pleases him more than money: justice.

What's It Really Worth?

"I've got a million-dollar idea" is a common expression. People who are able to convert what's in their mind into a million dollars in the bank are far less common.

The reason is fundamental to the business side of invention: nobody wants an invention per se; everybody wants an opportunity to make money. This is true whether you are an outsider seeking to sell or license an invention to a corporation or an employee seeking to free your idea from the corporate morass.

So you can think of the value of an invention as the ascending side of a bell curve. There is a continuum between a naked idea or an undeveloped paper patent and a finished, ready-to-market patent. As the best embodiment of the invention is developed, as durability or stability is proved, as cost of goods and manufacturing investment are quantified, as customer acceptance is demonstrated, and as a strong patent with a clean patent file is issued, the value of the invention rises step by step.

Wherever you are along this bell curve, sooner or later you'll find yourself negotiating with a corporate executive who has mixed emotions. This executive wants to put your product into his development and marketing chain and make big profits for his company, but he doesn't want to make a mistake. Most especially, he doesn't want to make a big mistake, because heads roll with big mistakes, whereas little mistakes (under $100,000) can usually be buried quietly somewhere.

You, on the other hand, want to get as much money up front as possible. For one thing, you'll be surrendering control over the fate of your brainchild to an unknown corporate maw. If the troops mess up and you get the invention back, you may have real trouble selling it again. In addition, you'll remember reading in this book that the payment at signing must be enough to keep the project high on the company's list of priority projects and also to be an embarrassment to the people working on the project if they quit at the first sign of problems. How big this number should be is a function of (1) the "state of finish" of the final product, (2) the probable size of the market, and (3) how many companies are either lusting after or yawning at your invention.

Please note that I'm spotlighting what's called the down payment, the option price, the front money, or the signing bonus. Almost always, such a payment covers a fixed period of time—and, from the inventor's viewpoint, it should. It not only serves as a prod to action, it makes it possible to split the early payment into several parts keyed to the calendar or, if you have some control over the progress, to events. As the product comes closer to fruition, you can always extend option times or defer the next payments if you're satisfied that the company is chugging down the road to market launch.

If you can take an invention through the developmental phases and put it into a successful test market yourself, the immediate value leaps by an order of magnitude. If you start a company based upon the invention and keep it alive for a few years, you no longer have just an invention to sell but a technology-based going business—and then you negotiate for an acquisition, not just a patent license. Dick Berger and Jim McLaughlin did the test market number with Carpet Fresh and sold it to Airwick Industries. Later, Dick Berger and other associates created Oral Research, Inc., to market PLAX and sold the going company to Pfizer, a story told previously in this book.

Whose Invention Is It, Anyway?

Gould was fortunate in having had Goldmuntz to protect him when he worked for TRG. "Who owns the intellectual property?" becomes an especially complex question when the mind that gave birth to the idea belongs to a head that belongs to a body that "belongs" to a corporation.

There is an inherent tension between company management and any employee who may declare that a certain invention belongs to him or her personally. How this dichotomy is resolved depends upon many elements: What was the employee hired to do? What kind of employment document (if any) was signed? How does the invention relate to this person's job description? What company materials or equipment was used, on whose time did the creative acts occur, and what are the attitudes and willingness to fight of both management and employee? In addition, a small but growing number of states have laws that protect employees by narrowing the scope of the "invention assignments" a company can require a person to sign.

Here is the distilled essence of wisdom on the ownership of intellectual property from the employee's viewpoint: *Everything depends upon having a clear, signed, written document executed before the invention, not a goodwill negotiation afterward.*

I learned this in a very personal way a zillion years ago when I worked for RCA, writing puff pieces for Victor records. It occurred to me that record albums of musical shows and operas would be much more enjoyable if the listener could read either a full or an abbreviated text embellished with photographs, both to better understand the lyrics and to fill in the action between songs. So I invented the RecorDrama album. It was simply a regular album of 12-inch records with printed pages bound between each record sheath. But it was new enough to get a U.S. patent. And I was naive enough to sign a release of all rights in and to the invention for the honorary sum of one dollar. I probably got what I deserved because I was too dumb to ask for more.

Getting Millions From the Mind to the Bank

I know no more ideal example of the right way for an inventor in a corporate cocoon to protect himself than Harvey Phipard. It was he who invented a "simple" sheet metal screw while working for the Continental Screw Co. and turned that new twist into a million dollars a year for years on end.

A large man with the charm of W. C. Fields and a figure to match, Harvey Phipard's rustic appearance belies his Massachusetts Institute of Technology training and his razor-sharp mind. As described in Chapter 5, Phipard had, as a condition of his employment, reached an agreement with Continental that they would negotiate in good faith regarding whatever he might invent. And that's what happened on several minor inventions.

Then in 1960, Phipard invented the trilobular sheet metal screw. Suddenly, the company viewpoint changed. Phipard's boss set out to license third parties even while the patent was still pending. And Continental began production. But they told Phipard, "Oh no, that was invented on company time, so that belongs to the company." In fact, Phipard had conceived the idea at home one evening and doodled the design on a desk notepad and presented it to Continental as his invention. Samples and development work were then done at the plant.

As it slowly dawned on Phipard that this might be a valuable invention, he began pressing for a deal but without results. Then, according to Phipard,

> They got me in my boss's living room one night, and with the bridge lamp shining in my eyes, told me how ungrateful I was not to turn the patent over to the company. And they finally said, "Now look, we're going to be very generous to you—we'll give you $5,000 cash on the barrelhead right now if you turn over all your rights to this invention."

Phipard made the decision of a lifetime, one that all inventors might seek to emulate.

"No," he said firmly, "I won't do that because I don't know if it's worth that much and that would be unfair to the company. Or maybe it's worth a lot more. And that wouldn't be fair to me. Whatever it turns out to be, I just want to share part of it. I don't want to give it up."

Because life does not work out as simply and happily as Candide might have expected, Harvey Phipard had to sue Continental Screw. After two and a half years—during which time he remained with the company as an employee and a director—his lawyer, Charles Pender of Choate, Hall & Stewart, managed a settlement that gave Phipard his first big royalty stream. Phipard would receive a small annual payment for Continental's use of the patent plus one-third of all royalties collected from licensees in the United States and 50 percent of royalties collected from abroad. The deal was actually more complex, since Continental had, two years before the patent issued, licensed a competitor, the Parker-Kalon Company, and given it the right to sublicense. Continental got back only 40 percent of the royalties collected by Parker-Kalon and Phipard got his third of that.

That was not the end of Harvey Phipard's battles with his employer. Several years after the royalties began to flow, he was in the front office of Continental and, as he relates it,

> I heard one of the bookkeepers say, "In which account do I put this Conti money?"—that was the name they had for foreign royalties. And I said to myself, if there's more than one account the royalties go into, something funny's going on. It took me six weeks to get them to admit they had sold the foreign patents to this Conti outfit for only half of the

> royalties Conti collected. Anyway, they paid me all the back royalties. For two and a half years, I'd just been getting a royalty check. No explanation or anything. After that, they gave me a complete statement.

I cannot bear to leave the story of Harvey Phipard without mentioning the millions that he missed because of a classic good news/bad news sequence.

While still working for Continental Screw, he got to thinking about the Phillips screw—you know, the one that has a crisscross pattern on the top that tapers downward. Harvey saw a weakness in that product: When you twisted the screwdriver with much force, it would often slide upward and out of the screwhead.

Harvey designed the "Phipard Recess." It was a rectangular slot with straight-driving faces. The driver didn't back off when you tightened the screw down. It worked.

Patents were applied for and—good news—granted. Came the bad news: Continental, because it was a primary licensee of Phillips, decided not to commercialize the system.

Then a bit of good news. While Phipard was burning over Continental's inaction, time went by and the Phillips patents ran out. Phillips Screw Company came to Continental and said, "We'd like to promote the Phipard Recess in our company line." Then came the really bad news: Continental's patent attorney, who by then was in his eighties, had forgotten to make the required payments to keep the patent in effect in foreign countries.

"So we started making these screws in this country and Canada, where the patent was still in effect because there were no maintenance fees then," Harvey told me, "but they were never a great success because they were more expensive. But I think practically all of the recess-type screws in Europe are now made with this invention and I don't get a penny in foreign royalties."

How High Is Up?

Starting and running a company is more difficult than licensing an idea; it demands a greater variety of skills and more allies and more money. But it's often the road to maximum money because a going company is worth more than the right to practice a patent. It's the way to find out how high the price can go for intellectual property, once the doubt has been taken out.

Say it's about 1984 and an engineer brings you an invention that he's been working on for years. He's all worked up about it, and he tells you the world really needs it. "It" turns out to be an electric toothbrush.

Ugh! First of all, Squibb has had the Broxydent electric toothbrush out there for years with its up and down action, the Bass method. Then there have been half a dozen entries with rotary action and multiple tips, and they all were big bombs. The only product to make it big in oral hygiene appliances was WaterPik, marketed by Teledyne. Furthermore, the inventor has been shopping this electric toothbrush around, trying to license it for eight years. A polite "no," right?

Not if you are Steven Lindseth, a twenty-five-year-old toiling in the vineyards of venture capital deals. Steve listened to the inventor, George Clemens, explain how he had watched the twisting of the lines on sailboats and noticed how they lengthened and shortened as they twisted. Clemens applied the principle to making a toothbrush with two rows of counterrotating multifilament bristles. The bristles, he claimed, would run in between the teeth, especially at the gum line. And it would rub off plaque much better than would a conventional toothbrush or any other electric toothbrush.

Steve Lindseth was intrigued. He took many months to visit dentists and dental professors, seeking to determine if Clemens's concept might make sense to professionals. He interviewed consumers too. Finally, convinced that he had a potential winner, he and his father, Jon Lindseth, a highly successful entrepreneur with a number of metalworking businesses, decided to launch the electric toothbrush, the one you now know as Interplak.

Because Steven had already been working in the venture capital field, he felt confident that the necessary capital could be raised, but as Jon told me,

> When we went to the venture capital community and said we had a $99 electric toothbrush, they looked at us as if we were loony. Steven knew all the venture capitalists in the United States. He sent a brief description of this thing to over 100 of them. We met with twenty-seven of them. And we were turned down by everybody. We knew in our hearts we were right, but we didn't have the analytical data. Our gut feeling said, if we could help solve the problem of periodontal disease, all you'd have to do is talk with one person who had gone through periodontal surgery and you'd find they'd willingly spend a hundred dollars to avoid going through that agony."

Fortunately, Jon Lindseth's businesses generated excess cash. So he personally invested half a million dollars to start the company, an amount that grew to $3 million before outside financing came along. Ultimately, to move the company to the point where it could self-fund its spectacular growth, $6.5 million of equity captial and $2.5 million of debt were required.

Money alone, however, was not the key to success. The Lindseths understood that they were not equipped to run the company, so they reached out for top-drawer personnel. And this was their first wise move.

As Jon Lindseth described it to me, "When I licensed the patent from the inventor, my job was to go out and find the team. I hired a first-rate headhunter to look for extraordinarily talented people, experienced people in the particular business we're talking about." The first three executives recruited for Interplak were paradigms of the policy: John Trenary, who left Teledyne WaterPik to become VP engineering and later president of Interplak, and Hugh Shores and David Landry, both from the Kerr Company, a maker of professional dental products, who became VP sales and VP professional relations, respectively.

It was not your ordinary hiring deal, however, as Lindseth took pains to point out:

> We also make our executives invest money in the deal. They've got to cough up cash. We want them to mortgage their house, we want them to be really committed. And if they are, they work hard.
>
> I've got a house out in Vail and last Saturday a friend was over visiting us and I was talking to our guys out in San Diego* and my friend said, "You mean they're in on Saturday?" and I said, "What do you mean *in on Saturday?*" Everybody's there Saturday and most of them will be in on Sunday. The reason they are there is that they own a piece of the action, and if it's a great success, they are going to make a lot of money.
>
> To give you an example, the president and a vice-president of our toothbrush company each put up $25,000 and in four years made about three million bucks.

* At Thermoscan, Inc., another Lindseth start-up, detailed in Chapter 13.

Another Round of Invention

With financing and key personnel in place, the Lindseths had two major areas left to attack: the manufacturing process and marketing.

As the product design was refined and manufacturing specifications spelled out, it became apparent that there were serious sourcing problems.

The bristles had to be made in clumps of filaments and be cone-shaped at the top, a shape related to their pattern when spinning. Toothbrushes are a bear of a product to manufacture and there are only a handful of makers in the United States. Interplak went to them all, without luck. They looked at cutting them with lasers, cutting them with water. Then they designed and built their own machine, a decision that was an added risk but also an added layer of protection against competition.

Each brush head has two rows of bristles, each with five multifil ament shafts that rotate counter to each other. All of the previous electric toothbrushes had all the bristles moving in the same direction; it was this counterrotation that enabled Interplak to get a patent with that capability as a basic claim. "A very tough claim," as Jon Lindseth put it.

To create this counterrotation, the Interplak brush requires ten tiny gears inside the brush head, driven by a shaft that stems from the handle of the brush, where the motor is housed. These tiny gears take a beating. They reverse forty times a second and are exposed to the saliva of the mouth and to dentifrice abrasives. No domestic factory would make the gears, but Interplak's people found a source in (no surprise) Switzerland and one in (surprise!) Communist China. Each made the gears perfectly to spec.

Even so, the original model broke down often enough to create an unacceptable return problem, and the product was redesigned to make use of a replaceable head. One nice part of the return headache was described by Jon Lindseth like this: "We got people calling up and saying to us, 'You bastards, my toothbrush doesn't work, get it fixed immediately, I love it!'"

In light of the problem in sourcing gears and the wave of early model returns, the second wise move was mandatory: They instituted a program to constantly upgrade the product, specifically by imposing tight manufacturing standards and quality controls at their Far Eastern suppliers' plants.

Solving a Marketing Dilemma

If wrestling with manufacturing problems was a tough chore, solving a marketing puzzle on which many had previously floundered demanded yet another wise move. The problem had three, highly interrelated components: pricing, distribution, and consumer acceptance.

Although Jon and Steven Lindseth and the Interplak team had confidence that the consumer would pay $99 for a brush that would help protect a person's teeth, and although the cost of goods dictated that price if start-up costs were to be absorbed, it was clear that the marketplace offered no such assurance. Other electric toothbrushes sold for $39 to $69—when they sold at all.

The road to consumer acceptance, they decided, had to start with professional acceptance. So they went first to the practicing dentist, the one person who could understand the way that Interplak cleaned interproximally and subgingivally (between the teeth and under the gums) and who could identify patients who needed this device. To the dentist, recommending a $99 item that would benefit suffering patients would be good medical practice.

So they launched an intense campaign to reach the dental profession. They enlisted the counsel of Dr. Walter Cohen, former dean of the University of Pennsylvania Dental School and editor of *The Compendium of Continuing Education in Dentistry.* They underwrote a large number of clinical studies in various parts of the country and sponsored a symposium at which ten of the clinicians presented their independent studies. It was, in effect, "instant peer review." Then, Interplak published the results of all ten studies in a special supplement to the *Compendium,* bought a 50,000 overrun of the publication, and mailed it to dentists who were not subscribers.

That was only one of many continuing efforts to persuade dentists that Interplak was an ethical, professional instrument. Incidentally, no mass consumer advertising was used at this introductory stage; but, because the product was both new and visual, it made good print and TV stories, appearing on both the "Today" and "Good Morning, America" shows.

The results were superb. Of the 80,000 private-practice dentists in the United States, Interplak eventually had 15,000 of them actually selling the device to their patients. What about the other 65,000 dentists? There was a plan for them, too.

Jon Lindseth described it to me as something for everybody:

> If a dentist believed in it but said, "OK, but I don't sell appliances and stuff—I fix teeth—but I'll tell the Jones

> Pharmacy, Mr. Jones on the first floor of my building, that I recommend it," then we'd say, "OK, we're going down and tell Mr. Jones that you are going to recommend it so he'll have it in stock." Then we taught the pharmacist how much money he could make by putting Interplak on the drugstore's display stand close to the cash register.
>
> We didn't stop there. We created an elaborate mailing program whereby, if the dentist would let us, we'd send literature to all his patients. We even had a coupon program that included an opportunity for the dentist or pharmacist to get a free Interplak for his wife or whatever.

There was one serious problem with the professional approach to creating a market: It was slow. The sheer mass of work—calling on dentists, attending meetings, waiting for clinicals, and so forth—cost Interplak the most precious asset of all: time. In addition, early product breakdown problems created a backlash from dentists who did not want to hurt their relationships with patients.

So after building an initial base of professional acceptance and a strong presence among independent druggists, Interplak sought to expand its availability. *The Sharper Image*, armed with the clinical results that made possible hard-hitting mail order copy, featured Interplak in its catalog. And Interplak became one of their biggest-selling products of all time, a reflection of what good advertising can do for a product that needs explaining.

The major hurdle for Interplak before it could reach its potential was, of course, chain distribution. The major drug chains and mass merchandisers could be expected to double the dentist-referred independent store business. But chain buyers considered the $99 price outlandish and, even more vehemently, objected to the $66 price at which Interplak sold to them. That problem was solved by getting the product into test stores to prove that it earned its shelf space in terms of turnover.

Jon Lindseth told me that the chairman of Walgreen's had stated at an annual meeting several years ago that this toothbrush changed his view of merchandising in America. Explained Lindseth:

> Walgreen's never thought that they could sell much of anything priced over $7.95 on the South Side of Chicago. Yet they sold this $99 toothbrush, two to five of them a week. Since retailers use gross margin per square foot of shelf space as the measure of success, Interplak, which scarcely occupied one square foot, was clearly returning

more dollars for the space it occupied than products with larger markup. Interplak was earning for Walgreen's in a week the same dollars other products were producing, per square foot, in a year.

So the marketing strategy was to start by setting the price high, gaining professional acceptance, driving for limited "core" distribution, and then moving out to mass distribution.

Granting that Interplak entered the market at a time when there was growing public interest in removing plaque and preventing gingival disease—fanned by major advertising by P&G and Colgate for tartar control toothpastes and plaque-removing mouthwashes—it was a prize case of turning an invention into a fortune.

Interplak did $4.5 million worth of business the first year and lost a million. The second year, it did $23 million and made one million. By the third year, sales were $65 million and profits $8 million, pretax. In Year Four, Interplak broke the $100-million barrier with humongous profits—$20 million.

In that fourth year, 1988, Jon and Steven Lindseth and their investment partners and the stockholder-managers of Interplak sold the company to Bausch & Lomb for $133 million—all cash.

Was the Interplak success just a freak, one of those wonderful experiences that reinforces your belief in karma, or was it a series of exactly right moves that can be repeated? Read about Thermoscan in Chapter 13 and draw your own conclusions.

What Are the Right Questions?

As one very capable invention manager said to me, "The way to buy assurance in the invention business is to ask the right questions as early as possible in the game. And if you don't get useful answers, change the questions—or if you're sure you've got the right questions, find some new people to talk to."

Whether you are an inventor or an entrepreneur or a corporate manager, here are some starter questions to help you evaluate an invention's potential, to consider whether to take it to market or to license it or to walk away from it altogether:

1. *How much inherent power does the invention have to excite people?* It's what I refer to in Chapter 8 as instant impact. When people see it, do they say, "Wow!"—or do they ask, "Why?" It doesn't take a major research project to answer this. Just make sure you're

showing your innovation to the right people, the kind who would be real, prime prospects.

2. *How big is the market in which the new product will compete?* Is the market growing or slowing? What is the likelihood that your innovation will expand the total category? Is there a preexisting demand for an improved product or do you have to convince people that they have a problem and that your invention will solve it? (That's called consumer education, a very expensive process.) Such marketing questions are fundamental, easy to ask and often hard to answer, but without at least a rough cut at them, it hardly pays to pursue your invention.

3. *To which businesspeople, in addition to the inventor and his or her family, is this new idea an important business opportunity?* If you're looking for licensees, how many companies are likely to consider this a major opportunity? Also, what company would be hurt by your product's introduction, and how can you turn that to your advantage?

If you're going to make and sell the product yourself, what are the problems of getting distribution, that is, how many potential dealers and/or distributors will jump at the chance to sell your product?

4. *What third-party forces are likely to help you?* Are there any professional groups, public interest groups, or other consumer groups or suppliers or makers of other products that will benefit from your presence in the market? In sum, who will support or cheer the new product without your paying them?

5. *What is the very best embodiment of or application for this invention?* "Best" in this case doesn't mean elegant or spectacular but simply the fastest, most surefooted way of getting the invention into a commercial mode and starting a cash flow. When the Diamond Fever Gang pursued the process for making synthetic diamonds at General Electric, some hoped for gemstones, but the big money was in tiny granules for abrasives.

6. *How long a time and how many dollars will it take before the invention reaches fruition?* It's a rule of thumb that getting from a paper invention to a going business will take twice the time and twice the money that anyone, especially the inventor, ever expected.

7. *How secure is the new idea?* And for how long? You can judge this from an independent patent attorney's opinion, an independent engineer's opinion of the know-how, a marketing expert's assessment of the value of the trademark, the size of the market, and so on. The critical business judgment, however, rests upon how much time and

money will be needed to bring the invention to a positive cash flow—because the ultimate security against competition is to be entrenched in the marketplace.

From Gould, Phipard, and Lindseth, three vastly different cases, you can see that there are many roads to converting intellectual property into cash—and not a few detours.

The universal headline, common to each of these stories, is this: By definition, an invention, a new-new product, has many unforeseeable problems—from patents to production to promotion—and successful inventors/entrepreneurs focus from Day One on determining the core problems and finding the answers to them.

To deal with issues and to capitalize on your intellectual property, you need only three things: money, guts, and all the allies you can find.

Chapter 7

Handling Foreign Rights . . . and Wrongs

You can generate as much money from an invention from the rest of the world as you do from the United States—if you manage it skillfully.

That's a good rule of thumb, but, like all generalizations, it is an averaging of winners and losers and it conveniently skips over the time, effort, and cost of getting it to happen. Not to mention whether or not your invention "fits" in various foreign countries.

Nonetheless, there's major money to be made from exporting technology or products, too much to ignore.

Ideally, a really worthwhile new product should be launched worldwide in one spectacular introduction. That preempts competition, avoids being dependent on the business cycle in any one country, and provides economies of scale early on. However, it isn't easy on executives and it demands much more capital at the outset than building country by country does.

One person who did this daring feat is Gérard Hascoët, who in the mid-1980s put together Technomed International to make and market a lithotripsy machine named Sonolith®. In the lithotripsy process, the patient lies in a pool of water while focused ultrasound shock waves are hurled at gallstones or kidney stones, causing them to disintegrate within minutes. Seeking to commercialize a lithotripter licensed from France's National Institute of Health and Medical Research, Hascoët convinced a group of investors that the only way to cash in on this radical replacement for surgery and to stay ahead of the field was to start a truly international business *tout à coup*, as the French would say.

Hascoët saw basic advantages to justify his ambitious plan for Sonolith: Lithotripsy was a dramatic breakthrough, reducing the patient's pain and risk without reducing the surgeon's income; no competitor had a real time advantage; and the Technomed machine was priced far below competing equipment.

Two factors, however, made a single-country start-up dicey. The largest and richest markets—the United States, Japan, and Germany—were the most heavily regulated, making approval time for a new procedure unpredictable. And big competitors like Siemens, A.G., were moving into the field of lithotripsy; clearly, if one waited, it would be hard to sell hospitals where the other guy's equipment was already in place.

Gérard Hascoët structured Technomed with an internationally staffed headquarters in France and operating subsidiaries in the United States, Japan, Germany, and Italy—the last-named being a land where regulatory approval was available for the asking, assuring early in-hospital experience. He divided the rest of the world for coverage by regional managers until additional subsidiaries were justified by growth.

From its ground-zero start-up in 1985 with only $5.5 million in capital, Technomed International was in the black by the end of its second fiscal year, and by 1991 sales were running at $60 million a year. Hospitals and governments in more than thirty countries had bought the Sonolith.

Not everything, however, went according to plan. In the United States, where the potential was the largest, FDA approval was delayed and clinical data submitted by a competitor were questioned, casting doubt on the efficacy of the process.

In addition, a new surgical procedure called laparoscopic cholecystectomy was introduced to the market. This is a procedure for reaching the gallbladder through the patient's navel by opening two or three 11-millimeter "ports" and then, using advanced video technology, grasping the offending organ with a little clamp and pulling it out. Surgeons could remain happily within their domain, even though they had to learn a new procedure, and lithotripsy seemed doomed.

As Jean-Luc Boulnois, Technomed's executive vice-president, pointed out to me, it was fortunate that Technomed was multinational from the start. While approval in the United States was hung up, sales of the million-dollar Sonolith in other countries sustained the company, allowing preparation time for a new product: a pulp-dye laser lithotripter named Pulse-o-Lith.

In real life, it turned out, one out of ten patients with gallstones also had tiny stones or fragments lodged in the common bile duct below the gallbladder. This required a second surgical procedure, often a second visit to the operating room. The pulp-dye laser lithotripter made it relatively simple to ensure a clean bile duct at the same time that the laparoscopic cholecystectomy was performed. Within a year of its introduction, Technomed International was selling, worldwide, about fifty Pulse-o-Lith units a year at something over a quarter of a million dollars each.

Having once survived the specter of being wiped out by a competing technology, Technomed management moved promptly to put other revolutionary products into its pipeline.

Jean-Luc Boulnois described the most spectacular of these like this:

> I think we have a unique situation here. It's a thermal therapy for treating benign swelling of the prostate gland, a common problem for men over fifty. Technomed has developed a proprietary miniaturized catheter, a transurethral microwave treatment. It's well known that hyperthermia will reduce prostate swelling, but it will also create fistulas in the rectum wall. Technomed's system generates heat only at the end of the microwave antenna, it incorporates a cooling system, and it has real-time thermometry to monitor what is happening. At $700,000 per unit, we have sold forty-five units in one year worldwide—and that doesn't include the United States, where we have clinicals in progress at many prestigious hospitals.
>
> We have a follow-up product in development, too—it uses a technique called photo-dynamic therapy—a state-of-the-art technology for laser surgery. And we got it by acquiring a division of a large European company.

Why did I give you so much detail about Technomed's products? Because it gives texture to Technomed's business strategy: Spread the market penetration risk around the world and have the next product ready regardless of how the first one fares.

Around the World in Eighty Ways

Almost all large corporations are committed, at least on paper, to doing business worldwide; just note how many signs outside corporate edifices read "World Headquarters." Inventors/entrepreneurs can't ignore foreign markets, either. Without foreign patents, without a drive to enter foreign markets, soon enough some determined Japanese or German or Taiwanese or Italian will market an identical twin in those markets, preempting the opportunity. Worse still, the competing product may also start appearing in the United States—backed by a very aggressive pricing policy.

So it's not a question of *whether* to push out around the world to every country where the technology or product fits. It's a matter of *when:* When can you afford it, when must foreign patents or trademarks be filed, when will you have the manpower to follow through?

The first thing you have to know about marketing abroad is that Wendell Willkie was wrong when he talked about "One World." There are many "worlds" out there. They have different languages, different customs, different custom duties, different profit structures, different distribution channels, and different patent laws.

Some small examples: In Japan, your patent application will be made public in eighteen months, but it will probably take five or six more years before your patent will issue—and it will cost you a bucket of yen. If you file in South Africa, your application will be "published" in about six months, long before it's been examined, giving everyone around the globe a chance to think their way around it. If you file in Latin America, you're probably wasting your money because it's impossible to enforce patents and next to impossible to get royalties out of most countries.

Because it is complex, you need to plan your international exploitation very early on. It affects everything: how the product is designed, the size, the color, the label, the trademark, the materials used—everything. Here is one small but perfect example: When Thermoscan, the in-the-ear thermometer, was being designed, its makers found that it was simple to provide a little switch to flip from Fahrenheit to centigrade, thus avoiding a separate model and a separate inventory for foreign sales.

If you expect to get back more money than you expend in foreign marketing of your invention/product, you can focus your decision making by digging out answers to these three questions:

1. *Where and when are you going to file for patents, trademarks, and copyrights?*

2. *How are you going to generate foreign income—by exporting the product or by licensing the invention?* If the answer is "exporting," will it be through distributors or a joint venture or sales representatives, and how are you going to find the right people? If the answer is "licensing," how will you find the best licensee, how do you negotiate the "right" terms, and how do you police the operation?

3. *Who is going to implement your foreign plan?*

It would, of course, take an entire book or graduate school curriculum to answer these questions in depth. So what follows is more inspirational than educational, more replete with insights than complete with data—and therefore more useful to the inventor/entrepreneur.

The International Patent Maze

Whether and where to seek foreign protection is a decision best made when you know, first, that you have something worth protecting—in other words, a winner—and, second, that it has a reasonable chance of succeeding abroad as well.

At least, that's the theory. In practice, you're almost certain to come up against decision deadlines before your knowledge of foreign opportunities is complete. The most urgent of these decisions concerns filing for patents, trademarks, and copyrights. In order to keep your U.S. date of filing as your priority date abroad, you must file a foreign patent application in most countries within one year after filing in the United States; otherwise, it's the date of your patent, country by country. Furthermore, in some countries, such as Germany, you must file trademark applications within six months after filing elsewhere. So your invention is quickly open to the world for knockoffs, making it important to complete both patent and trademark filings wherever you want coverage.

Foreign patent applications are costly. Filing fees, your lawyers' fees, the foreign correspondent lawyers' fees mount up. And when you get the patent, there are maintenance fees. So deciding where to file is a key decision. The go/no go test is whether or not a given country is either a large market for your product or a logical manufacturer of such products. The alternative to patenting for countries where the product life cycle is short or the dollars involved are minimal is to cream that market and save the patent maintenance fees.

Trademarks and copyrights are a somewhat different case. The costs involved are not substantial if you have any serious intention of selling in a given country. And sometimes it's easier to combat trademark misuse than patent infringement.

With the advent of the European Economic Union, filing for intellectual property protection in multiple countries becomes more attractive. The EEU intellectual property system will take some time to shake out. Meanwhile, multicountry filings can be done under the

European Patent Convention (EPC) or under the Patent Cooperation Treaty (PCT).

Under the International Copyright Convention, if the notice is properly displayed, copyright protection is afforded in over sixty countries. And under the Paris Convention, signed by ninety countries, a trademark owner is entitled to six months after filing in one country to file in any other.

For a good briefing in this area, read *Copyrights, Patents & Trademarks Worldwide* by Hoyt L. Barber. Never, however, be your own lawyer, and, most especially, never in the field of international proprietary rights. Not only are they complex and occasionally devious, but the governing bodies keep changing the rules. The European Community, even as this is written, is about to propose new international rules to protect intellectual property. Not unexpectedly, the United States, Great Britain, Germany, and other high-tech countries have pushed for stringent rules on everything, including computer software, while the less industrialized nations are for short patent life and freedom of reverse engineering.

So knowing how things stand and where they are going in patent, trademark, and copyright law around the world is an arcane specialty within the legal profession. Not every patent and trademark law firm is up to speed, and you don't want them to be studying the subject on your time sheet. Don't be afraid to ask for credentials or to bring in a co-counsel for international work.

Partnering Abroad

Licensing an invention abroad and exporting a manufactured product are two totally different situations that happen to have some things in common.

The principal difference is, well, patent. Most licensors have much less power over their licensees than exporters have over their distributors. It can be difficult to know if you are getting an accurate count of sales, if your product is receiving a fair share of marketing effort, or if your licensee is busy trying to invent around your patents. Later in this chapter, I'll be specific about some ways to guard against such disappointments.

The chief commonality is that, whether exporting or licensing, you need not merely a customer but a partner. Almost without exception, the people who consistently make money internationally do so by establishing long-term relationships with their foreign business partners. This cannot be done by correspondence or telephone. Here are a few cases to illustrate the right way to go.

When David Montague was still in the early stages of success in the United States with his BiFrame bicycle, he received considerable favorable publicity, including a story smack-dab on page 3 of *Newseek*'s international edition. As he explains it,

> We had perhaps fifty different companies around the world calling us and writing to us, but even if they say they want to buy some bikes, you just can't sell them right away. Selling a few here and there just doesn't do it. The important thing is to establish a long-term relationship. I think the thing to do is to get over there and deal directly with the distributor, see their facilities, and check their distribution network. You have to talk with people in the industry and find out what kind of reputation they have.

Nor did David use any intermediaries. He believes that reps or brokers just add another layer of cost and he believes it's faster to "just do it" than to search for reps through trade sources, the U.S. Department of Commerce, or through the commercial attachés of various countries. I know people who have used both these routes to find overseas customers, but the success of Montague BiFrame Bicycles, a small company with a unique product, certainly proves that old-fashioned American directness can work with the right product.

When I asked David how he knew which companies to go after, he replied,

> Basically, I just went to the major trade shows in each country and saw who was big—you know, who had the big booths. In France, you start by talking with Peugeot and in the U.K. you start with Raleigh. Then you go down the line to the next biggest and the next. As you start talking with people, you get an idea of not only who is big but who's good and who's coming up. You set your priorities and look for a commitment. We're in nine countries now, including England and France and Japan and Australia—and I did it that way in every one of them.

Herb Allen developed European business for his Hallen Company's marvelous wine bottle uncorker, the Screwpull, by capitalizing on contacts he had made in pursuing his own personal interest in fine wines. For many years, Allen traveled in Europe with his wife, visiting wine merchants and wineries and shipping hundreds of cases of his selections back to his own wine cellar and to friends. When he was ready to sell Screwpull abroad, Allen supplied samples to some of the most prestigious wine merchants in Europe and demonstrated it at wine tastings. Such wine aficionados hailed the product for its ease of use and for ending the era of the crumbled cork. The resulting publicity in print and by word of mouth created product credibility that gave Allen a strong hand in securing distribution.

With endorsement by the cognoscenti and with per capita wine consumption increasing worldwide, Herb Allen was able to sign up the best distributors from Europe to the Far East. Then Hallen Company made a major commitment and bought a factory in England, providing entry into the Common Market without import duties. The results, according to Ann Mannix, president of Hallen, was that business in Europe and Canada became twice that of the already healthy U.S. business.

In international trade, there are scenarios for dealing with such items as import quotas, import duties, withholding taxes, and currency fluctuations. Getting the right overseas distribution deal depends on how you cope with such economic realities.

One creative technique for exporting patented products was related to me by Michael Lyden, vice-president for corporate development at Tyco Industries. Tyco is America's fifth-largest toy company (as of 1991), with hits such as Typhoon Hovercraft and Quints and Magna Doodle. Mike told me,

> To avoid pricing ourselves out of foreign markets, we sell to our overseas distributors at a very low cost. If we sold them at a full export price, the import duties—always a percentage of declared value—would raise their costs still further and then they have to put their markup on top of that. Our products would be dead on the counters.
>
> However, because we have patented products, we can and do charge the distributor a royalty on sales. We are licensed by the inventor, and the distributor becomes a sublicensee of ours. With variations, it works out in major markets around the world.

You may wonder how the inventor makes out in such an arrangement. The Tyco answer is to pay the inventor the toy industry's customary 5 percent royalty on sales through their owned-and-operated overseas companies but only 3½ percent on sales through distributors. This is fair to the inventor, Mike Lyden pointed out, because the price of toys overseas is normally higher than it is in the United States, so the net dollars of royalty are about equal. It's also plain that it adds 1½ percent to Tyco's kitty when negotiating with overseas distributors.

Exporting the Right Stuff

How successful U.S. companies are in exporting inventions in the form of finished products is a matter of great concern to those who would like to leave their children a country as prosperous as the one they enjoyed. And it is all too obvious that when it comes to VCRs and other electronic gear, American ingenuity in design and manufacturing economics have taken a beating from the Japanese and their Pacific Rim copycats. Yet there are exceptions that demand attention, possible paradigms for a revival of America's reputation for innovation, not to mention its balance of trade.

One such case is what the Edgecraft Corporation accomplished with Chef's Choice, the first knife sharpener that produces professional results at home the first time you use it and every time you use it. So it meets the initial criterion for export success: The product actually works better than anything out there. It also meets the second test: It has meaningful patents in every significant country where it could be made or sold.

Both Dan Friel, inventor of Chef's Choice and founder of Edgecraft, and some of his associates had previous experience in foreign markets. They went country by country to locate the most capable distributors. Within a few years after the American launch, they had locked up deals with outstanding housewares people in the United Kingdom, France, Germany, Switzerland, Italy, and the Scandinavian countries. Concurrently, Edgecraft moved to the Far East, to Japan, Singapore, Hong Kong, Australia, and New Zealand.

But even with overseas marketing experience in other products, how did Friel and friends know how to find and choose the right distributors for a product like Chef's Choice? They used all the standard techniques, and the operative word is *all*. As Dan Friel described it,

> You go to the leading department stores and ask, "Who are your leading distributors? Who works best with you? Who sells the upscale products?" Or you go to the annual Domotechnica show in Cologne and open up a little booth

> or you walk the floor to see who would be the likely distributors and talk with them. Sometimes you get one distributor to recommend another. You get out there and get very active and keep digging and following all the leads. Weeding out and selecting, making a few mistakes, trying to avoid big ones.
>
> Mostly, we lock into companies for the long term. Occasionally, we have to make a change. In Japan, for instance, we went with Dodwell Company, a big importer. They import Royal Dalton China and the Prestige Cooker and some other things that made them seem right for us. They also happen to be big whiskey importers; they're the chief Scotch importers in Japan. Recently, they decided there was more money in consumable liquids than in long-lasting housewares. so we're going through the process again, talking with fifteen or twenty distributors there until we find the right one.

Anyone taking a new-new product to such lion's dens as Hong Kong or Japan must consider the possibility that they are simply inviting early imitation. However, the risk is less than it might appear. In most fields, the copyists from the Far East are continually scouting the U.S. market for products to knock off, so it's wise to go after business wherever you can as soon as you can.

Edgecraft developed a good, tight patent position and filed in countries that were either primary markets or likely imitators. When I asked Dan Friel why he sells in Singapore without patent protection, he said,

> You know, that's really not a big enough market to attract someone to copying. We are exposed in those minor countries. You just can't cover all the bases. But we do have patent coverage in Korea, Taiwan, and Hong Kong. And we've tried to go into those markets and saturate whatever market there is before anyone else can get active.

All in all, by being both aggressive and quick to move into foreign markets, Edgecraft was able to establish itself as the controlling factor in electric knife sharpeners, even though the price abroad was slightly above the U.S. norm.

Dan Friel offered wisdom for many inventor/entrepreneurs when he said, "Even if you're not large, you need to be a worldwide company... you need to be internationally minded. Provided you can.

It isn't always possible. But if you don't move at an early date, they'll freeze you out."

To which I would add, don't assume that you *can't* sell abroad without plenty of hard thinking. Case in point: Would you expect to be able to export potato chips? Of course not. Big volumes to ship, low unit prices. Well, how about Canada? Oh yes, that's an export market. Stephen Bernard, founder of Cape Cod Potato Chips, thought about it, and foreign sales became 15 percent of their business. They hooked up with Murphy Potato Chips, a well-established, top-drawer distributor north of the border, and as Bernard says, "Murphy did a tremendous job in Toronto, Montreal, and the eastern provinces of New Brunswick and Nova Scotia. We truck it to them in Canada. And we also send Cape Cod Chips to the Caribbean by boat. If shipping costs weren't so high, we'd send them to Europe."

What made it possible for a commodity like potato chips to succeed internationally? A substantially differentiated product "invented" by automating a chip-cooking process that had been abandoned fifty years before. Shades of Saratoga Chips!

Licensing Abroad

Licensing an invention in foreign countries involves all the same problems as exporting—finding the right partner, making the right deal, and so on—with one critical addition. You must identify someone with either a terrific imagination or a tremendous need—or both. You're looking for a *fit*—a company in which your invention/product is exactly right for the kind of thing it likes to do or, from a market standpoint, needs to do.

Of course, it helps a lot if your invention is already being marketed successfully in the United States; that provides some credibility and a cash flow to finance a foreign licensing program, a not-to-be-underestimated point because foreign licensing deals almost always require repeated visits to the prospective licensee plus expensive patent filings and maintenance fees.

What all this implies is homework. Here are the fundamental questions to ask yourself before making an initial decision to enter a foreign licensing program:

1. *How strong is my proprietary position?*

Clues: Even more than in the United States, make sure your patent attorney is getting you a first-class search of foreign patents and publications.

Remember, in your own country you have the advantage of

knowing what's going on in your field through professional journals, conventions, and trade gossip. Before you spend the money to launch a foreign licensing program, especially if you are going for real gold, get someone to do the homework in major potential markets. Such digging can tell you not only whether your patent position is likely to be strong but also what competition your potential licensee will face.

2. *What is the receptivity level for my new-new product/ invention in the largest foreign markets?*

Clues: This is, in the first place, a sociomarketing question. Are people in a given country new product–minded? Will you have to change people's habits to win acceptance? Can these people afford to change to your product/technology? Is the marketing and promotion infrastructure available to jump-start an introduction?

Second, are there qualified prospective licensees who are open to outside invention, evidenced by what they have previously done, not just by what they say?

3. *Who are my best licensing prospects and how can I determine who the key players are and establish entrée at that level?*

Clues: Licensing abroad cannot be done by mail, phone, or fax. Nor are deals concluded with a single visit. So it's almost always cheaper and more effective to spend the time and money needed to zero in on the best situations. Rehearse before you go overseas and once more before the meeting, make sure you get the decision makers, and have a spare of everything from the projector lamp to your product prototypes. Avoid making solo presentations unless you're one-on-one with the top person at your target company.

Also request an *advance* list of everyone who will attend the meeting from their side together with the title and/or role of each person in the decision chain. One job for someone on your side is to determine during the meeting who on their side is the likely deal killer and who the likely champion, and then quietly go to work on that latter person during lunch or dinner.*

4. *How can I hedge my bets?*

Clue: Don't depend on only one potential licensee for a given territory until the deal is signed and the check clears the bank.

What's So Different Over There?

All the preceding clues are useful for translating principles into

*In some countries, this is normal; in others, it's a no-no. You just have to know where you are.

actions. Some insights, however, are easier to glean from specifics than from generalities. So here are some case-in-pointers.

If you are doing your own licensing abroad, put real effort into both identifying the people you want to meet and securing introductions to them.

When I was looking for a European licensee for Valentine Laboratories' unique antacid tablet, one that turns into liquid as you chew it, my first step was to obtain through friends at an international advertising agency the names of the market leaders in England, France, and Germany. I checked these leaders against names supplied by the New York commercial attachés for those countries—obtained by simple phone calls. Then I asked a vice-president at my bank to query its branch operations in each country as to contacts with my target companies. I also asked the Pharmaceutical Manufacturers Association in Washington to give me the names of the appropriate person in the corresponding organizations in each country and then telephoned these people for advice (call at 6 A.M.—it's less expensive and gets a European's full attention.)

Then I called three licensing officers with whom I had done business previously at U.S. pharmaceutical companies and asked them for any personal introductions they might make or, failing that, the identification of who really counted in my target companies.

One result was that England became the obvious first stop. Per capita consumption of antacid tablets was highest there and British brands were beginning to penetrate the continent too. From James Church, senior vice-president of Marion Laboratories and a man who had made friends throughout the world, and from Adrian Huns, a marketing manager at Boots, the largest British drug chain, came the identical recommendation: See Tony Jamison, who runs the Nicholas-Kiwi Company, maker of Rennies antacid. Both Church and Huns encouraged me to use their names for introduction.

Nicholas-Kiwi licensed the patents not only for England but also for many countries on the continent and for Australia, the company's point of origin. Naturally, there were problems in scaling up the production and questions about selecting flavors. But Tony Jamison wanted the product to win and he told his team to find solutions, not excuses. When the Nicholas-Kiwi people were satisfied that they could make and sell the quick-liquefying antacid tablet, they also took a license for a calcium supplement tablet. Tony Jamison became a good business friend and went on to become the head of all European operations when Sara Lee Industries bought Nicholas-Kiwi.

No matter what path *you* may follow to foreign licensing, it has to aim at the same target I have just described: the right man in the right company.

Pay heavy attention to local customs and business facts. When you're on other people's turf, seeking to license your innovation for their territory, you can build respect for everything else you will say by showing sensitivity to their mores and markets.

One inescapable starting point is language. If you do not speak the language of the country you are visiting, at least learn how to say, "How do you do? I'm pleased to meet you. I'm sorry I do not speak your language. May we talk in English?" The fact that English is now the chosen language of international business and science does not excuse you from making that effort.

Nor does it counter this advice to engage someone fluent in the local tongue to be part of your presentation and negotiation team—both to explain any points that your audience may not grasp (even though they speak English) and to hear their reactions and sidebar commentary expressed in the language of the land.

Business habits are another area worth your attention. In some countries, it is gauche to discuss business during the lunch or dinner break but de rigeur to imbibe much wine. As a consequence, you may be less prepared for the ensuing business talks than your host. Unless you're well practiced in this drill, drink a glass of buttermilk before the meal or protest that your ulcer is burning up. For insight into the importance of local customs, see Figure 2.

Knowing local practice in terms of distribution patterns, profit structures, and so on is equally critical. Knowing your prospect's competition and past strategies is most helpful. And knowledge of local purchasing habits is indispensable. For example, in the United Kingdom the largest-selling antiperspirant is Ban® Roll-on, a product of Bristol-Myers; and the reason it outsells others is that customers can buy a refill bottle that just screws onto the old roller-ball cap. So it's not only a pretty good product, it's cheaper than competition. That makes it tough to license a new brand, no matter how good it may be. You have to add to new product cost not only a royalty for the invention but the cost of a non-refillable package and the high costs of market introduction.

In many parts of Europe, foods such as mustard, jellies, and pâté are sold in tubes; in the United States such products would lie on the shelves unrecognized. In France, suppositories are the delivery system of choice for a wide range of medicines; in the United States, they're practically a form of punishment.

Whatever the local customs and local market options, you have to

(Text continued on page 140.)

Figure 2. Do's and don'ts of gift giving abroad.

What gift do you bring to your business host on that next overseas trip? Should it be expensive? What about a family gift? Here are some tips to keep you out of trouble.

Japan

- January 1 and mid-June holiday seasons are special times for gift giving.
- Don't expect recipients of your gift to be effusive in expressing appreciation. They may not even open it in front of you.
- Brand-name items are appreciated.
- Let your counterpart initiate gift exchange unless you are the host or your gift is a reciprocation.
- Your return gift need not reflect 100 percent reciprocity. Thoughtfulness is more important.
- It's not unusual to receive a gift when you first meet a Japanese business associate. You need not immediately reciprocate with a gift. If you will feel uncomfortable without one, however, it is wise to be prepared.
- Don't out-gift the Japanese. Gift giving is more their custom than ours. Allow the Japanese to derive satisfaction from giving and avoid obligating them by giving a more expensive gift.
- Present a gift when the recipient is alone, unless you have gifts for everyone.
- People usually like consumable gifts and small conversation pieces, i.e., liquor, candy, cakes, and books.
- In general, Scotch is preferred to bourbon.
- Always bring a gift when visiting a home.
- Logo items should be unique, but not a joke. The logo should be subtle.
- Wrap non-logo gifts. Avoid bold colors, dark gray, black, or black and white combinations. The black-and-white combination is reserved for funerals.
- People like to get flowers. The 15-petal chrysanthemum is acceptable, but 16-petal chrysanthemums are in the imperial family's crest and should not be used commercially.
- Don't open a gift in front of your host unless you ask whether your host would like you to do so.
- Don't give surprise gifts. The receiver may be embarrassed by not having one for you at the moment.
- Children love the latest American toys.

- Don't make a ceremony of presentation. It should seem spontaneous and sincere but never a source of pride to the giver.
- Whenever possible, give a gift that shows that you did your homework. Take time to know your recipient's personal preferences.
- Entertainment should be of high quality. However, you need not match the royal treatment typically provided for visiting business associates by the Japanese.
- Avoid ribbons and bows in gift wrapping. Our bows are considered unattractive. Various colors of ribbons have different meanings. Rice paper for wrapping signifies good taste.
- The color red implies happiness and good health. Use it.
- Don't offer a gift depicting a fox or badger. The fox is the symbol of fertility; the badger, of cunning.

The Arab World

- Don't bring liquor as a gift. Liquor is taboo in the Islamic religion.
- Avoid junk gifts.
- It is often rewarding to give something with intellectual value, such as a book. This will compliment the recipient's concept of an educated self.
- Don't give a gift on first introduction. It may be interpreted as a bribe. If you give a logo gift, be sure it is unique or has some special significance.
- Bring gifts for children but *never* for a wife or wives.
- Don't let it appear that you wanted to present the gift when the recipient was alone. It looks bad unless you know him well. Give the gift in front of others in less personal relationships.
- Something to be used in the office is usually acceptable.
- Consider carefully the nature of the relationship when selecting a gift.
- American and German merchandise are considered good-quality gifts.
- Be careful when selecting items depicting animals or animal sculptures. Many connote bad luck.
- Don't admire an object openly; you may be the recipient of it.

Europe

Manners are vital in France, West Germany, and Great Britain. Actions speak louder than words. The person who knows the proper rituals for every occasion will be in good stead with most European businesspeople.

(continued)

Figure 2. *(Continued.)*

- Always send flowers before arriving at someone's home for dinner. It is acceptable to bring them with you, but preferable to send them.
- Avoid red roses and white flowers, even numbers, and the number thirteen. Don't wrap the flowers in paper.
- Don't bring perfume to a woman unless she has requested that you purchase a certain type for her.
- Silver items are appropriate gifts if you are hosted at a German home for a few days.
- People appreciate fancy chocolates and special liqueurs.
- Don't risk an impression of bribery by spending too much on a gift.
- Gifts for children are especially appropriate.
- Enclose a card containing a hand-written sentiment with personal and courtesy gifts.
- Gifts should be simply wrapped.
- Avoid logo gifts, unless they are unique.
- Europeans appreciate porcelain gifts. However, home decorations are risky unless you know the decor preferences of the recipient.
- Personally engraved writing instruments or sets are well received.
- Gifts with historic or intellectual appeal are especially appropriate.
- There is no substitute for local advice. Always seek advice of a native resident or the local U.S. consulate when in doubt. The value of a gift should be based not on money but on its appeal to the individual recipient.

Latin America

- Until you develop a personal relationship, do not give a gift. An exception is if you want to show appreciation for hospitality.
- The best time to present a gift is after business negotiations have been completed.
- Never go empty-handed to visit a home.
- Avoid the colors black and purple. Both are associated with the Catholic Lenten season.
- The number 13 is considered unlucky in most of Latin America.
- Do not give a knife (implies cutting off a relationship) or handkerchief (associated with tears).
- Latin Americans often appreciate small electrical appliances from the United States. However, it is best to ask if there is something that you might bring. This will avoid implying that U.S. products are superior.
- Something for the entire family, such as artwork from North America, is a good idea.

- The latest toys from the United States will please children and parents.
- Thoughtfulness counts more than monetary value.
- Give gifts during social encounters, not during business. Since business is not conducted during lunch, this is often a good time to present a small gift.
- Logos on gifts should be subtle, unique, or have some connection with the representative's company other than that implied by the logo alone.
- Women executives should be aware that gifts to men may be misinterpreted. Don't be paranoid; just make sure your well-intended gift is accurately interpreted.
- Find out what items are unavailable or heavily taxed. If you bring a gift that is locally taxed and pay the tax, the gift will be appreciated more.

The People's Republic of China

- Don't give clocks. The pronunciation of the word *clock* in Chinese means "funeral."
- The Chinese appreciate small mementos.
- Don't give lavish gifts.
- Never make an issue of a gift presentation, publicly or privately.
- Don't give American currency. It is illegal to bring in outside currency.
- Gifts should be presented privately, except collective ceremonial gifts at banquets.
- Don't present gifts until all business negotiations have been completed.
- The Chinese appreciate gifts associated with your home state or company.
- Kitchen gadgets, photobooks, name plaques for desk use, personally engraved pens, records, or a good brandy or cognac all make nice, inexpensive, useful gifts.
- Become familiar with the legal implications of expensive gifts.
- Don't wrap gifts before passing through customs. They will be unwrapped there.
- People like simple wrapping or a tastefully decorative gift box for courtesy gifts.
- Familiarize yourself with Chinese banquet entertainment customs.
- Always have a good reason for a gift, one that the recipient can use to justify accepting the gift.

Source: Adapted from "International Business Gift Giving Customs" with permission from Parker Pen Company and Roger Axtell.

know as much as possible in advance and subtly let your prospective licensee know that you know.

This is particularly serious when it comes to business-dealing practices. For example, it is a common experience among American businessmen who have dealt with the Japanese in particular but also with Koreans, Taiwanese, and other Far Easterners to come away from an initial presentation with the impression that it was a great success and that a deal is imminent. This rarely happens. When things go well in such meetings, what I have heard is a sharp sucking in of breath through the teeth—a restrained Nipponese form of "Oh, wow!" —followed by copious taking of notes in Japanese. When the Japanese say *hai!* they don't mean, "I agree with you," they mean, "I hear you."

At the end of the endless questions, when you ask for some level of commitment or at least response, you are told that they are very impressed but must report to their superior. A few weeks later, you are told that they still have "a few follow-up questions" and "could you please send some more samples?"

Now that is just one of many negotiating tactics that teach the wisdom of keeping secret as much as you can, even though you are selling. You can be glad-handed anywhere in the world, including the United States, only to find that the next meeting the company holds centers on how to beat your patent. The problem abroad is that it's harder to analyze what is going on and to keep posted.

Picking Partners Market by Market

One key to coping with this puzzle is to find a licensee who really needs you and to set the deal where your continuing services or flow of technology are important to that licensee. Another is to find a qualified person to represent you, a person who knows the market and the licensee and who can be there regularly.

Whatever path you follow, remember that the objective is to make the licensee your partner, not your customer. In mutual interest lies safety—even with the Japanese. Then write a tough contract.

An obvious and fair question after reading the above would be, "Why not just license my development to an American company and let that outfit sell it through its subsidiaries or partners overseas?"

The answer is a qualified "yes." One-stop shopping always sounds ideal. The fact is, however, that very few American companies have a strong presence in *all* the major markets of the world. Gillette does.

Caterpillar does. Some drug companies do, but only for specific product categories, such as SmithKline Beecham's position with Tagamet.

So negotiating an international deal with an American company should go like this: "Let's settle the U.S.A., O.K.? Then let's talk specifics about your market position and selling capability market by market." I learned this the hard way. Early on, I would normally offer a license for the United States and Canada as a package. Then I found out that, even with powerful companies, the manager of the Canadian subsidiary may have his own plans which leave no room for a new development from the home office. And the best companies let their best managers make these local decisions. So why give away a territory in advance of determining whether or not the local management will exploit the opportunity? All this advice applies equally to an individual or to a company manager seeking foreign licensees.

Most large European and Japanese companies think of the world as their oyster, but they rarely have strength in all major markets any more than American companies do. But they'll take as much territory as you give them in the hope that they can either use the product to bolster their weaker territories or sublicense it or use it as part of an exchange of rights, if your license agreement allows that.

Do It Right or They'll Do You Wrong

Doing business abroad isn't just different from doing business at home. It's tougher—because you don't know all the rules and because arm's-length transactions are different from ocean's-width transactions.

But it's an enriching experience, literally and figuratively. So, as I said at the beginning of this chapter, if your invention or product is suitable, you must sail forth to find foreign gold.

As a *bon voyage* present, I give you some of the hidden rocks to watch out for.

Financing Exports

Assuming yours is a young company without all the money in the world, how are you going to finance your production from the time you order the raw materials until the end product is accepted in Japip? Of course, you're going to get an irrevocable letter of credit from your customer because collecting even from creditworthy foreign customers typically takes more than a hundred days.

Will your bank advance you enough money against the letter of credit to operate? And at what rate? Try for a commitment from your bank *before* you go chasing orders around the world; check overseas banks with U.S. branches. Work with your banks to establish standards for the kinds of orders and paperwork that are acceptable for them to advance a certain amount of cash. Will your suppliers help by holding their bills until the transaction is completed? Can an export sales agency finance the transaction in return for a fee?

Of course, every exporter faces such problems. But your company may be able to do something about it if your product is new-new instead of me-too. In return for a hefty cash advance, you can offer your overseas customers exclusive distribution rights forever, or for a few years. You can negotiate for prepayment with a cash-rich customer in return for an anticipation discount. Or will the U.S. Export-Import Bank help you with foreign credit insurance or with a working capital guarantee? How about the Department of Commerce and the Small Business Administration in Washington or special export-help programs that a few states run? All these sources keep changing what they will do, but the fact that you are exporting American innovation can give you the extra edge to get through the official jungle and help you to find someone with a real heart who can get you what you need.

That's hardly a complete list of what is needed for success in exporting. You have to cope with currency fluctuations, import duties, foreign taxes, extreme heat or cold that can destroy products, thievery on the dock, and much more. It should at least put the thought in your mind that financial arrangements are the *sine qua non* of foreign business.

Licensing Arrangements

In addition to the safeguards detailed in Chapter 11, there are both legal and business issues to watch for in foreign license agreements. Here's a sampler:

- *What currency do you get paid in?* My preference is always for being paid in U.S. dollars. True, the dollar goes up and the dollar goes down, but it's better to plan your future without trying to outguess the financial markets—or the convertibility of a foreign currency.
- *How can you avoid collection problems?* You do this by building advance annual or quarterly payments into the agreement. Always try for an arrangement that puts enough money in your pocket so that if your licensee should stop paying royalties or stop making the

product or otherwise be in breach of the agreement you can send a termination notice and use their money for legal fees to seek whatever is due you.

• *What should you do about "improvements?"* Any sophisticated licensee will ask that rights to any improvements you may make be automatically included in the agreement at no additional cost. In some cases you can argue about the cost-free part, but the key response is that you will in turn be free to use without cost any improvements that the licensee may make and also to include such improvements in your licenses to third parties in noncompeting territories.

• *How can your protect yourself against double-crossing engineering?* There is no preemptive language your lawyer can put into an agreement to prevent your licensee from having his creative geniuses engineer or design around your patents and know-how. This is where the business side of invention takes over. You must set it up so that it is in your licensee's business interest not to upset your relationship. For instance, it can be done by a cross-license of the licensee's technology, or by offering a first refusal on your next invention, or by stipulating a balloon payment if the licensee terminates or defaults on the agreement. Or get the chairman's youngest child admitted to medical school at your alma mater. Whatever. Just remember, while all is still sweetness and light as the license agreement is being negotiated, find some significant bait to deter your licensee from even thinking about creating bitterness and acrimony later by trying to get around your rights.

Interpreting the Language of Foreigners

Whether you export or license, remember that communication is always a problem. Even though English is the language of the scientific world and, more and more, the language of the business world as well, correctly interpreting what is being said and being understood yourself is the key to success.

At Product Resources International a few years ago, a Japanese company met with us in New York, received a full presentation on our liquid fiber laxative, and went away politely proclaiming that it would be ah-so well-received in Tokyo. Here's a fax that came back to us many weeks later:

> Thank you for your sending information on your new technology liquid fiber technology. After an internal review on

> it, we'll let you know our comments. As you know, Nemoto-san is very much interested in this pharmaceutical form.
>
> I hope your business is going well. My business also keep me busy, so please give me a little time before we make comments on this matter.
>
> Let me allow to introduce my colleague who is working with me. His name is U. Shinoda, and he is very smart so, if I have a chance to take him to U.S.A. I would like to introduce him to you, and of course, when I visit New York next time, I'll call you and would like to invite you to the dinner at the restaurant "Top of the Window" which is one of the place I want to be.
>
> That is all for now.

You can interpret this message in two ways: as an indication of serious continuing interest in the technology, or as a not-too-subtle way of organizing an expensive dinner at Windows on the World, atop New York's World Trade Center. The right answer lies in understanding the behavior and style of the other person, not in his or her use of the English language. I recognized the message as a sample of what Hawaiians call *hoomalimali*, which roughly translates to "kid 'em along."

On the other side of the coin, the need to make yourself understandable to foreigners is at least as important as understanding them—and just as subject to communication vagaries. Especially because English is so frequently used in international business, there is a tendency for Americans to assume that others understand them perfectly when there is no blank stare or overt objection to suggest otherwise. You must be sensitive to the other person's desire to appear knowledgeable and go the extra mile to make sure you are understood. As we used to say in the Signal Corps, "Any message that *can* be misunderstood *will* be misunderstood." And cultural differences accentuate this possibility, or Winston Churchill would never have said, wittily, "The British and the Americans are two peoples separated by a common language."

Deciding Who Is the Best Man

Throughout this chapter, I have talked of a mysterious person: You. I have said you should do this, you should do that. What I mean is that you should get this or that accomplished, not necessarily do it yourself.

Earlier, Dan Friel and David Montague recounted their experiences in going to countries themselves and locating the right distributors. John Trenary, faced with developing an international market for Interplak while the product was still in its early growth period in the United States, took a slightly different route.

As John summarized his experience,

> I remembered a guy who worked for me when I was in the International Division at Teledyne WaterPik and I tracked him. He had started a company in Germany in the dental business, built it to about $20 million, and semiretired. I felt we needed a strong day-to-day presence in overseas operations, so we made a deal whereby he would be our direct representative in Germany and coordinate day-to-day activities in Europe under an agreed marketing plan for each country.
>
> We tried to use the same tactics as in the United States—that is, generate third-party performance data and have our distributor promote the product through the dentists. Here's what happened.
>
> In Germany, the plan worked well, excepting that we had to hire our own reps to call on the dentists and explain the product.
>
> In Italy, our distributor was wired into almost all the *farmacias* and felt we didn't need the dentists' endorsement. They got immediate penetration, and it sat on the shelves. We immediately went back and hired two direct people to call on dentists and began attending all the conventions—the product promptly started to move.
>
> In England, we literally had a product that did not fit because most bathrooms in England do not have an electrical outlet by the washbasin. We simply redoubled our efforts through the dental profession. We completed some very good local clinicals and persuaded dentists to put pressure on their patients to use Interplak, even if they had to carry it into the bathroom.
>
> In Japan, it is quite different. Products are sold both by the dentist and at retail—but you can't go both ways. We started with a dental products distributor and they did a good job. We got some sales and some professional endorsement. But by the time we sold Interplak to Bausch & Lomb, it was clear we would have to go to the retail side for volume and we moved that way, knowing that the professional side would pretty much collapse.

In the case of Interplak, John Trenary absolutely needed his own trusted representative to install and supervise the program. Other products, such as Chef's Choice knife sharpener or the Montague BiFrame bicycle, have no such requirement. But initiating any overseas distribution demands that working relationships be developed, and that cannot be done while you sit in a U.S. office.

* * * * * *

It's a big world out there and one of the first rules of multinational marketing is that each market is different. Generalities are useful only if you are keenly tuned to the exceptions.

Nevertheless, if you have a unique product that fits foreign markets—and you have reasonably good proprietary rights and the finances to implement overseas sales—and if you would like to be around in the twenty-first century, there is little choice but to strike out abroad. As the old sailing ship captains used to say to the sailors when a brewing storm demanded that they go aloft to furl the canvas, "Growl you may, but go you must."

Part III

Product Concepts: Finding Pathways to Market

Chapter 8

New Product Marketing Insights—Part I

While diligently searching for insights into new product marketing success, I visited the Baker Library at the Harvard Business School to see what their chosen books had to say about the subject. As I worked my way through row after row of five-level steel shelves crammed with perhaps a thousand books on marketing, turning the lights on and off in each aisle as the signs requested, I realized that I could never read all those volumes—and probably no one else had either. But I was impressed that the subject of marketing had attracted so many scholars and writers.

So I systematically pulled out every tenth book and glanced at the table of contents and read a few pages. They were good, mostly, explaining everything you always wanted to know about gathering market data, staffing a marketing department, managing a sales force, doing consumer research, and on and on. After checking about fifty books, it dawned on me that almost every author's point of view was based on having an established budget, usually a hefty one, and a staff of people to get the job done.

This is, of course, just the opposite of what most inventors and entrepreneurs have to work with, and even the new product champion inside an established company is sometimes shortchanged on budget and support personnel.

That's why you'll find this chapter light on conventional marketing methods. Instead, I've concentrated on the ways that bright people get products to market with less-than-complete data and less-than-desired sales forces and certainly less-than-the-hoped-for money. What I seek to do is simply to stimulate you to think unconventionally about how to launch a genuinely new product.

While the winners discussed here are a disparate crew, people who combined art and science, luck and skill, there are marketing insights in each case that you're not likely to find in a B-school library. If you find conventional marketing wisdom in this chapter, it's because someone applied it in an unconventional way.

To keep your marketing eye trained on the target, please take this as gospel: people don't really buy inventions. What they buy are technology-based new products—products in which the technology creates a benefit that they can see, feel, touch, smell, taste—or products that hit them in the pocketbook.

Even an invention that does nothing more than reduce the product cost for the manufacturer can provide a marketing advantage, too, if that extra margin is used to increase the promotion budget or to reduce the price to the consumer; these too are things people feel.

Spotting the Instant Winners

How do the people who make it big with their new-new products get early signals of success or failure? They deal with basics.

Nothing is more basic than people shouting "Hooray!"—literally or metaphorically—upon seeing the product or upon trying it for the first time. I call this effect instant impact. It's that flash of recognition that comes with seeing something you know you've always wanted. That's different from someone just saying politely, "Well, I believe I really did prefer product A to product B in my two-week home use test."

If your new product has only subtle or long-term benefits, you should seek ways to create instant impact, even if it is only symbolic. Particularly with truly new products, quick recognition of the benefit is the difference between success with a smaller marketing budget or failure with a larger one, between educating the buyers on a slow track and having the buyers boost your sales with free word-of-mouth recommendation.

When Ralph Sarich developed his small, light-weight automobile engine, he installed it in a Ford Cortina and took a picture of a technician working on the engine while standing alongside the motor inside the engine bay. He created instant impact with that picture because almost every automobile owner already knows that, with a conventional engine, there is hardly enough room for a mouse under the hood of a car.

Another example of the importance of instant recognition of a product advantage occurred recently in the shampoo business. Procter & Gamble was sitting with a tired old brand named Pert®, which had only a 2 percent share of the market. Then P&G researchers invented a new formula that combined a shampoo and a conditioner in one product. Such a combination was, in fact, an old gambit, an oft-used claim among shampoos; but those other products did not live up to their claims. The shampoo ingredients washed out the conditioners along with the dirt. The new Pert formula was dramatically different. As soon as you used it and dried your hair, you could tell that this one product did as much for your hair as using a typical shampoo and conditioner separately.

So even though P&G put a very modest advertising budget behind Pert Plus in test markets, sales doubled within six months and the product went on to become the number one shampoo in the country.

In case you're wondering if this success wasn't just a result of product positioning or great sloganeering, note this from the *Wall Street Journal*: "The success of a Pert Plus demonstrates one of the oldest marketing truisms: Consumers flock to true innovations, regardless of slick advertising or fancy packages."

Spotting the Natural Opportunity

Are there natural elements to look for that create instant impact, elements in the invention that tap into the desires of the buyer?

Absolutely. Especially if you can tap into preexisting, pent-up demand. You'd think such strong desires would be easy to identify, but this is not always so. When parents first saw the Ansa bottle with its center divided so that even infants could hold it, there were immediate cheers. But if you had asked people back in Radio Days if they were longing for television, you'd have gotten a yawn. Or asked heavy consumers of beer, circa 1960, how to improve their brew, they wouldn't have dreamed of telling you to go invent a light beer.

Even a long-standing, genuine dissatisfaction with a current product may be masked by resigned acceptance of the product's flaw—and revealed only when a solution is demonstrated. Shoelaces were taken as an inescapable fact of life, even though they took time to thread through the eyelets and even though the laces came untied, until some unsung hero invented loafers. Staying with the feet for a moment, it's said that women's high heels were invented by a girl who was kissed on the forehead. In short, people don't usually talk about—or even admit to the existence of—a problem until you show up with a solution.

For both inventors and company managements, it pays to direct efforts toward inventing products that evoke the response, "It's about time somebody did that!" To identify such target products, the best technique I know is to ask every imaginative person you know to complete the question: "What if we could———?"

A second natural element surfaces when the product dramatically solves the problem that it addresses. It delivers what it promises so quickly, so effectively, so much better than what has gone before that people say, "Aaaaah!"

A New Twist on Corkscrews

Herb Allen, for example, invented the Screwpull®. If you drink wine, you may well be familiar with it. Screwpull is a revolutionary corkscrew. What makes it different is that the cork climbs up the spiral screw rather than having to be muscled out by sheer force or mechanical leverage applied by the user.

Now what led Herb Allen, the well-to-do retired president of Cameron Iron Works, a leading supplier of oil-well drilling equipment, to invent a new corkscrew and go into the corkscrew business? For starters, he had traveled extensively in Europe and had learned to appreciate good wines. At the same time, unable to turn off the keen curiosity about how things work that had led to his inventions of blowout protectors for oil wells, he had observed many, many different kinds of cork pullers. Some wouldn't center, some chewed up the cork, some required great strength, and some just didn't work at all. Herb found dozens of cork-remover inventions—not only variations on the classic awl type but such amazing or amusing inventions as a Swiss uncorker that uses an air pump with a stainless steel needle to inject compressed air into the bottle and thus eject the cork; a German gadget called Ah-So with two narrow blades that one must insert along either side of the cork and then twist and pull out; and even a motorized corkscrew.

Now Herb Allen was not a casual observer of life's scene. He found that the first corkscrews dated back to the seventeenth century and that although people had tried to make corks twist and climb up a spiral screw, no one had designed a functional corkscrew that did not depend upon powerful vertical force to tug the little bugger from the bottle.

He joined the International Correspondents of Corkscrew Addicts, a self-selecting group of wine lovers fascinated by the challenge of opening wine bottles. As he traveled from country to country, he talked with people not only about the wine they shared but about the way they popped their corks.

Nobody thought their corkscrew was great. Interesting designs, yes, but functionally only *mezzo-mezzo*. Most important, Allen noted the intensity of resentment among wine lovers when a cork stoutly resisted their tugs or crumbled into their treasured fruit of the vine. Herb Allen, the curious observer, was doing more than tasting fine wine; he was doing excellent market research.

Gradually, there formed in his mind a set of requirements for the ideal cork remover. It should automatically center the screw over the cork. It should require minimum physical effort. It should do minimum damage to the cork itself. It should be so different from existing devices that it would provide a strong patent position. It should be easy for anyone, male or female, to use.

All these requirements fermented in Herb Allen's mind, but he did nothing with his ideas until he retired. Then one day, his wife, who liked to serve wine to her friends when they came to lunch but who hated opening the wine bottle, seeking to put her husband out of harm's way as her female friends arrived, said to him, "Dear, why don't you go down in your workshop and invent a corkscrew that even I can use?"

Embracing the project with the same systematic approach that had created his 150-plus patents in oil-field equipment, he attacked two key requirements for a breakthrough design: (1) the screw must start out perpendicular to the center of the cork, and (2) the cork should come out slowly because it takes less force to move a cork at low speed than it does at higher speed.

Over a two-year period, the core solution materialized: Drive a stainless steel, Teflon-coated spiral "worm" down into the cork, and when the downward movement of the worm is stopped by the plastic frame in which it is mounted, the cork will begin to rotate and climb upward on the Teflon-coated wire until it is out of the bottle. Allen designed the plastic frame to sit on the lip of the bottle and center the worm over the cork. A small plastic cross-bar "handle" atop the worm provided all the leverage needed, which wasn't much.

I've described Herb Allen's invention in some detail so that you can see how well it matches what he had observed people needed in a cork remover. When he demonstrated it for wine buyers, he got the immediate "Wow! Great!" response he expected. Mind you, these people had not been running up and down the streets demanding a new and better wine bottle opener. But when they saw one that was not only easy to use but didn't bollix up the cork, they toasted the inventor and said they were ready to buy. Where the market actually lay, however, and how Herb Allen reached it is related later in this chapter.

Turning a Nightmare Into a Business

Another example of a problem that no one had solved was how to treat babies with colic. Colic hits about 15 to 20 percent of babies between the ages of three weeks and three months, and occasionally up to six months. It hits the parents of those children even harder.

When colic strikes, a baby screams and cries and may appear to be in pain. The episodes are irregular but repeated. Doctors are unsure of the cause, although it may be digestive or emotional. What is certain is that colic is a nightmare for parents, who must walk the floor holding the child for hours on end, often all night, trying to maintain their emotional control.

In 1986, the trials of managing his child through colic attacks were visited upon Armando Cuervo, manager of international business for Ashland Oil Co. One night, after many difficult bouts with this problem, Cuervo called his pediatrician and begged him to open his office to see the child.

Cuervo bundled the screaming infant into his car, and a funny thing happened on the way to the medic: The baby calmed down and fell asleep.

Reflecting on this the next day, he realized that children frequently fall asleep in a moving car. He wondered if there were elements in that automobile experience that could be reproduced elsewhere, in a crib, for example.

Armando Cuervo went to General Motors, Ford, and Chrysler and asked all three what frequencies and amplitudes of sound and vibration a child would normally experience riding in a car at 55 miles per hour. The invention was already clear in his mind: a sound generator on the crib rail that would duplicate the lulling road noise of a moving automobile and a vibrator attached to the bottom of the crib to recreate a car's shaking motions. He called the product SleepTight® and named his company—what else?—SweetDreams, Inc.

A visit to his patent attorney reassured Cuervo on probable patentability, assuming that he could prove the product's efficacy. A patent search turned up only one prior invention: a mechanical hand that would pat the baby on the back—provided, of course, that the child didn't roll over and get pounded on the face.

One might think that a product which meets such a pressing, preexisting need by quieting that darling, bawling infant and letting you get some sleep would have instant impact and would rank near the top of the easy sell list.

Instant impact, yes; easy sell, no. Why? That brings up the other side of the new-new product coin: What are the fears, the concerns, the uncertainties about a new product that may make a prospect unwilling to buy it?

In the case of SleepTight, two obvious worries were the baby's safety and a $69.95 price tag for a device from an unknown company offering no absolute assurance that it would work. How Armando Cuervo overcame these sale-killers is an instructive example of the entrepreneurship involved in marketing.

Cuervo first tied up with a St. Louis company named First Photo, specialists in taking professional pictures of infants. In return for enclosing his brochure with the picture proofs sent back to parents, Cuervo gave First Photo a commission on sales generated through their leads.

"We got a very low percentage of returns," Cuervo told me. "I found out that parents were afraid to buy a medical device by mail order, and we had not yet nurtured the pediatricians to endorse us."

Cuervo's next move was to tie up with Ross Laboratories, a division of Abbott Laboratories. They had 600 salespeople in the field who called on every pediatrician in the country every two weeks to promote Simulac baby formula. To Cuervo's and to Ross Laboratories' surprise, most physicians did not want to be the sales channel for SleepTight. They liked the product when it was explained to them, but they wouldn't sell it. That meant promoting the product to the consumer, something Ross felt they could not do. Cuervo bought the business back.

By now, the marketing strategy was clear: Make sure you have the pediatrician's blessing, but sell SleepTight direct to the user. As Cuervo explains,

> Now we run small ads in such magazines as *Parents* and *American Baby*, and I still mail inserts through First Photo. But we also attend pediatric conventions and sponsor nurse education courses. When a parent calls us on 1-800-NO-COLIC, we have the professional background to close the sale. Now pediatricians accept us because they still have to deal with colic mainly by trial and error.

> Nobody really wants to give drugs to an infant. Now our sales are about 75 percent consumer and 25 percent pediatrician.

With sales of over 20,000 SleepTight units a year, Armando Cuervo is quite content with his name, SweetDreams, Inc.

Winners and Losers: Looking for Clues

Judging a truly new product in advance of actual market experience is difficult. The inventor, of course, is always ecstatic, but it's obviously essential to hear opinions about the brainchild from intelligent folks other than the "parent," and to do so sooner rather than later.

Putting aside formal market research for a moment, certain people have that special combination of native cunning and distilled experience that enables them to spot winners (and losers) early on. When the invention is sufficiently developed to show to a person knowledgeable in the field, it pays to have a confidential preview meeting. But with whom? Talk with a top sales rep in the field. Try the buyer for a big potential future customer or the editor of a leading trade magazine or the publisher of a newsletter in the field or the head of a trade association. Or find someone like Bob McMath of The Marketing Showplace, or Art Schwalb, publisher of *Marketing Intelligence*, or other people who have been around a certain class of business for years and who have developed a good sense of what will go and what won't. There is always a risk that such people may be wrong in their evaluation of your product's potential, but there's a good chance that their hard-nosed realism will spot any fatal flaw in the product or in the marketing scheme.

One way to develop that keen sense of judgment yourself is to look continually at innovations entering your fields of interest and try to determine why they succeed or fail. Even if you cannot always get finite answers, the exercise sharpens the mind as to the right questions. Then apply those tough questions to your own new products:

"Who is likely to jump for joy over this invention, anyway? Can I find a core market that will get me started fast? How aware, how angry are people about the problem this product solves? How fast can they dig the main benefits of this invention—or, conversely, how much do I have to educate them to understand it?"

Here is some food for thought as to what consumers go for or reject in new products:

Take Procter & Gamble's Pringles®, those potato chips that are so identically contoured that they come nested in a can. It was certainly an invention, a remarkable piece of food engineering.

The obvious advantage is that they don't break before you can get at them. The can kept the chips fresher than bagged potato chips of the era when Pringles was introduced. P&G's research must have told them that this was the greatest thing since sliced bread,* because they launched the product with a gigantic push.

But if you watch kids or adults eating potato chips, you'll see most of them stuffing all those bits and pieces into their mouths without hesitation. No matter what the market research may have said, it seemed that most people didn't feel very upset about broken potato chips—and, even more important, they were certainly not unhappy enough to accept the taste of the original Pringles product, which was a long way from the accepted standards set by Lay's or Wise. Mighty P&G kept Pringles on the market for many long years before it became profitable.

In contrast, as described earlier in this book, Steve Bernard's little company in Massachussetts broke into the field from scratch with Cape Cod Potato Chips. These chips were remarkably flavorful, not only when the bag was opened but many days later. It had a right to go into the business because it gave consumers unexpected, old-fashioned taste, made possible by incorporating an old-fashioned process in a modern production line.

Of course, every product category has its make-or-break hallmarks. In the food field, knowledgeable marketers will tell you that taste is critical to success. Thus, even with the proliferation of "good-for-you" products, the winners and the losers are separated by the consumer's perception of taste. Take potato chips again. When chip makers brought out "no salt" potato chips, they failed, because people thought the chips tasted yucky. But when the same manufacturers brought out "low salt" chips, there was just enough regular potato chip taste in them to make a tenable segment of the market (perhaps 5 percent) very happy.

Learning From Mistakes

Most of the time, the consumer can sort out real product improvements from imagined ones.

You can learn a lot about the importance of real product differ-

*A German baker was the first to slice and sell bread, but the first U.S. patent on a multislice machine went to Jay W. Currier of Springfield, Mass., on May 16, 1871.

ences by studying some "new products" launched by major marketers. Large corporations have so much marketing research and promotion power at their command that you might assume a major national launch could not fail. Not true.

It's especially illuminating to look at some not-so-new products with marginal performance differences—products that depend upon the halo of existing famous brand names or the genius of intrepid advertising men and women. Such new products, based primarily on the maker's avarice rather than on improving the consumer's lot, are frequently tough on the bottom line.

Clorox Co., for instance, decided to go after a piece of the massive laundry detergent business by marketing Clorox Super Detergent®, bucking P&G, Colgate-Palmolive, and Unilever. What advantage did this product bring to the consumer? Why, the name Clorox, of course. And what did that name suggest to the consumer? It suggested that there is Clorox bleach in with the detergent, right?

Now that may be a convenience sometimes, but do people want bleach in their detergent for their colored clothes too? No. Is there a different version of Clorox Super Detergent *without* bleach? Should the consumer keep on buying his or her regular detergent for colored clothes and Clorox Super Detergent for other laundry? Is the consumer getting confused? And what happens when P&G defends its franchise by bringing out a flanker product called Tide with Bleach?

Overall, what happens is that Clorox Super Detergent gets no more than 4 percent of the market at the height of its promotional investment, and that particular product introduction is reported to have cost Clorox some $45 million in fiscal 1990 and possibly twice that in fiscal 1991. P.S.: Tide with Bleach doesn't make a big splash either.

Meanwhile, P&G, not content to take a chance that Clorox Super Detergent might just succeed, launches another product that the world had not been clamoring for either, a liquid bleach called Lemon Fresh Comet®, aimed, of course, at undermining not only the original Clorox Bleach but Clorox's heritage and financial base. Lemon Fresh Comet Bleach died too.

There is an old business adage: Find a need and fill it. Just contrast the degree of preexisting consumer need as between Allen's Screwpull and Cuervo's SleepTight and Clorox Super Detergent or Lemon Fresh Comet Bleach.

Of course, they're in totally different fields. Of course, it's hard to be original in laundry products. But what about Bounce, the little sheets of fabric softener that you can throw right in with the load at the start of the cycle instead of coming back to dump in some liquid softener? Or the newer softener sheets that you can put in the dryer where the softening action doesn't go down the drain with the rinse water?

You can focus everything you need to know about picking new product winners and avoiding losers by remembering this rule: People won't buy anything the second time that didn't deliver on its promise the first time. And the best way to convince people that your product is different is to build in the kind of performance that users can readily recognize.

If that sounds too simple, remember what grandmother used to say about making better chicken soup: Put in more chicken. If the invention/product doesn't elicit excitement without promotion or advertising, be prepared either to improve the product or spend a lot of money marketing it.

Answering the Eagerness Equation Through Market Research

Assessing a new product's probable market size is perhaps the thorniest problem of all.

By definition, an invention is something genuinely new. To one degree or another, it is out of the range of people's prior experience. It works differently. It looks different. It may require the user to change his or her habits or beliefs. Therefore, prejudging how the new product will be accepted in the marketplace and deciding how best to introduce it and how to measure success or failure are challenging tasks. Later in this chapter, we'll look at some people who succeeded and some who failed to get over this hurdle.

As a practical matter, to decide whether or not an innovation is good enough to market, you have to analyze both sides of this equation: How dissatisfied are people with what they now buy versus how delighted they will be with the new product you offer? Think of it as "the eagerness equation."

In the final analysis—whether you are the vice-president of marketing for a giant company deciding to roll with a $50-million budget or the president of a start-up company deciding to launch your product with a $50,000 budget—you have to gather whatever facts you can on both sides of the eagerness equation and then make the go/no-go call based upon your own judgment.

Which brings us to the subject of *market research.* Evaluating new products, especially new-new products, by asking potential buyers how they like them is an arcane business. Various market research companies have their own "proprietary" methodologies and each company will assure you that theirs is the true faith. For example, National Analysts, Inc., of Princeton, New Jersey, has a method it calls Lifestyle Segmentation.

Alfred E. Goldman, retired president of National Analysts, explained the method to me this way:

> Different people expect different things from a given product. Different benefits of that product are more or less important to people depending upon their lifestyle, their values, their expectations. So we try to provide our clients with information about user attitudes that is arrayed so as to match their product's appeals against various lifestyle market segments, enabling them to sharpen their marketing focus.

The National Analysts system has value and so do many of the other specialized approaches of leading market research companies. To move into this kind of sophisticated research, you have to be ready to use the information when you get it. The Catch-22 is that it takes the income stream from a large going brand to support the cost of such research in the first place.

So for the inventor or for the executive seeking some bedrock information to lean upon when making early marketing decisions, here's a review of the most frequently used consumer research techniques and some comments on their value.

Focus Groups

A small number of people, say eight to ten, gather around a table with a moderator. The panel participants have been prescreened as prospects for the product. The moderator gently leads the panelists into a discussion of the product category and what they use now and what pleases or bugs them about a particular brand. Then the moderator introduces the product—either the real thing or a prototype or a concept drawing—usually with a prepared statement about the product.

Focus groups can produce useful insights into what people may be thinking about your product or how they react to existing products. You don't have enough people to take a meaningful yea or nay vote. Sometimes you have one or two members of the panel who dominate

Jacob Fraden works with a prototype of his Thermoscan infrared thermometer that will take a person's temperature in two seconds through the ear.

Jack Rabinow explains his invention for cutting granite quickly under water to famed sculptor Robert Berks.

Eugene Lang with three of the "kids" from P.S. 121 in Harlem where Lang underwrote college tuition for an entire class, an act made possible by royalties generated by his licensing activities.

David Montague holds the Schwinn Montague BiFrame, a full-size mountain bike, that folds in half for easy storage.

The late Herb Allen in his own wine cellar removes a cork from a rare bottle without effort using his ScrewPull invention.

Armando Cuervo with son Andrew whose colic attacks started Cuervo on the trail to his SleepTight invention.

Harvey Phipard, back in his days as an employee of Continental Screw, holds a giant model of his trilobular sheet metal screw.

Dan Friel with his Chef's Choice knife sharpener and a microscope he uses to check blade edges.

Richard Berger (seated, right) signs the $200 million contract with Pfizer for PLAX mouth rinse. Also seated (l. to r.) Dr. Allan Lazare and Truman Susman of Oral Research, Inc. Standing: two of Pfizer's lawyers.

Art Fry, inventor of 3-M Company's Post-it Note Pads, looks for new versions of the product.

Noel Zeller is wrapped up in his inventions, including the Itty Bitty Book Light (on the right).

Ralph Sarich developed this 90-lb., 1.2 liter Orbital engine that delivers 95 horsepower.

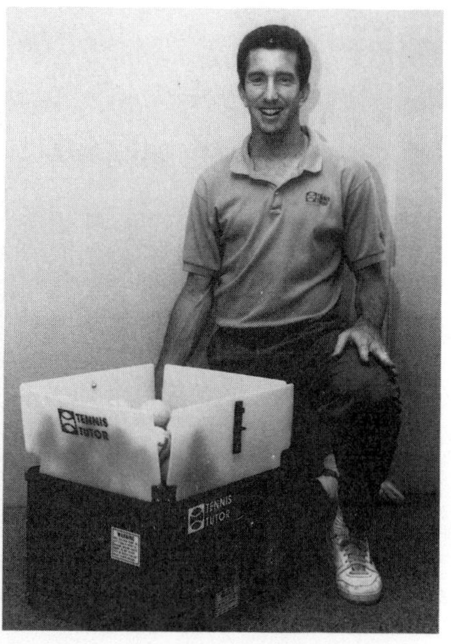

Bill Greene kneels beside his invention, the Tennis Tutor, the battery operated ball thrower that runs for four hours.

Dennis O'Connor holds Masco Corporation's Battery Buddy®, a monitoring/protective device licensed from inventor Jeffrey Sloan. O'Connor's face is framed by the Delta® faucet, the licensed product that fueled Masco's growth.

Marc Newkirk holds a Lanxide ceramic-reinforced aluminum disk brake rotor before a wall covered with plaques commemorating some of Lanxide's more than 100 issued U.S. patents.

Sanford Redmond squeezes mustard on his food from his dispenSRpak, using just one hand.

James Murdock comes up for air in an Endless Pool, his swim-in-place invention.

the others. Even though you warn them not to, panelists may be tempted to behave like "experts" rather than consumers. Because the new product is usually not presented in finished form and not presented with a finished advertising message, and because the panelists know they will not be asked to buy anything, the focus group is not a tool for testing invention-based products but rather a forum where high impact vs. low impact, love-points vs. hate-points can be sorted out by sensitive inventors and marketers.

What you can get out of focus groups depends almost entirely on who is listening to what the panelists say. If the moderator or the people behind the one-way mirror are truly careful listeners, they can pick up the ideas that plainly make sense, even if the ideas are expressed by only one or two panelists. And a group meeting is a wonderful, and sometimes the only, practical way to get a product in front of potential customers before the product design and package are frozen.

Wisely used, focus groups may be the one best chance to sort out strengths and fatal flaws early in the game. For an example of intelligent and fruitful use of focus groups, read the story of Dan Friel and Chef's Choice® in Chapter 12.

One-on-One Interviews

Going out to a shopping mall or to people's offices and collaring them for a person-to-person interview is a direct alternative to focus groups.

It's better in that the interviewer gets to look the interviewee in the eye for one to ten minutes and try to elicit thoughtful, candid answers. The subject isn't influenced by what other people are saying. And it usually provides a more random sample than you might get with folks who are willing to attend a focus group.

The downside is that one-on-one interviews cost more money; they largely depend on the skill of the interviewers; and the format is typically question-and-answer rather than discussion. And, unless you do additional research, you still don't have enough data on which to base a comfortable marketing action decision.

In any case, the true value still rests with the sensitivity of the interviewer. Some years ago, Dr. Syed Abdullah, an Indian-born market researcher who worked for my company in the early 1960s recounted his experience while doing one-on-one interviews for the Tea Council. The project, an effort to find out what you could say to people to get them to drink tea instead of coffee, was under the aegis of Dr. Ernest Dichter, a psychologist who became famous as a market researcher. Abdullah told me:

> At the bottom of my final report, I wrote a note to Dichter saying that people thought tea was an insipid drink for old ladies or for people sick in bed. Instead of being infuriated by this bad news, Dichter immediately tied it to the fact that most Americans didn't make tea strong enough because, unlike the English, they wouldn't wait for it to brew. He recommended a campaign to get people to brew strong tea, using words that would counter the sissy image. It came out this way: "Tea—Make It Hefty, Hale and Hearty." It got results.

Literally, of course, Dichter didn't invent a new product, but he did discover how to get consumers to "reinvent" tea in their own minds. And he left us a lesson as to the importance of listening and interpreting what people tell you about your product.

Telephone Interviews

This is essentially the same thing as one-on-one research, only worse. Compounding the reluctance of people to spend time on the phone with some unknown interviewer is the explosion of telemarketing calls, which has led people simply to hang up on strangers. Telephone research these days works best when you have a qualified list of people who have a reason to want to talk with you, a case in point being follow-up calls to check customer satisfaction with a product already purchased.

In-Use Placement

This is the basic tool for most mass-marketed products and for some industrial products too. It consists of locating a geographically and demographically representative group of users of the product being researched and then placing samples of the item with them for actual use.

The success or failure of this kind of research rests heavily on the market research organization doing the work. Key issues are (1) the size of the sample, (2) how reliably it represents the universe, (3) how the questionnaire is constructed and worded, (4) how the responses are supervised, and (5) how the statistics are interpreted when the answers are compiled. As you can surmise, this can result in a parlay of unintentional mistakes and misinterpretation.

Very well executed, however, in-use placement tests are a valuable tool. But they're quite costly, to the point where such testing must be weighed against a small actual market test in which you can get closer to how the customer actually behaves.

Which Crystal Ball Should You Choose?

For the inventor or entrepreneur or daring company executive, it's important to understand that most market research methods are geared to measuring relatively small differences between the existing market winners and the new product being tested. For instance, will a detergent with this new fragrance sell better than the same product with the present fragrance? Do you like the look of coffee blend A better than that of blend B? In many large product categories where the products themselves are very close in performance, such tiebreakers can make all the difference in sales.

But when you introduce a product that is out of the range of the customer's prior experience—say, a product like Max-Paks®, Maxwell House's premeasured coffee—you frequently can't go by what people say or even by what they do in a short test. The more innovative the product and the further it is from the prior experience of the customer, the more important it is to do a form of research that puts knowledgeable, insightful people close to the research while it's happening.

I had an illuminating conversation about market research for genuinely new and different products with Cal Hodock, past president of the American Marketing Association and senior vice-president of Comart/KLP, a marketing services company. He told me:

> For most relatively new products, especially mass market products, you can find some existing information, some bases for comparison, and make reasonable forecasts. Say you had a tennis racket that gave you an audio clue as to where the ball was hitting the strings, you could dig out enough existing data on tennis products to come up with something useful by translation. But if your product is the first...the first...a real first, well, you've got a real problem.

We talked about some cases. In particular, Cal pointed to Interplak, the electric toothbrush, a success story detailed in Chapter 6.

Based on firsthand study of that innovation, Cal pointed out that there was no reasonable way of forecasting this success by conventional consumer market research. Electric toothbrushes were a low-interest category. On top of that, the retail price of Interplak was pegged at $99, far above other electric toothbrushes and much more than retailers were getting for hair dryers and other small appliances. But Interplak had a wild card: Clinical studies showed that this toothbrush left teeth almost plaque-free, a definite contribution to reducing gum loss. With this factual tool, Interplak convinced dentists, and dentists told their patients what was good for them. To have interviewed consumers before that happened would have been to elicit a yawn; to have interviewed them afterward would only have confirmed that price was secondary to keeping one's teeth. Interplak took over half of the electric toothbrush market. Jon Lindseth and his associates sold the company to Bausch & Lomb for $133 million.

As Cal summed it up,

> That success would have been very difficult for a major company to forecast through conventional market research. The electric toothbrush market had been stagnant for years, and if you looked at the known factors, you would have shied away from that business. A major company would have to tool up the finished product and do both clinicals and a large home placement test before they would even go to test market. So it would never happen.
>
> Yet this product turned the category around and won an Edison Award from the American Marketing Association as one of the best products of 1988.

Despite the difficulty of researching new-new products, focus groups, when properly employed, can provide enough insights on new-new products to galvanize action toward marketing. But what does "properly employed" mean?

Dan Friel proved this with his Chef's Choice product. Like Interplak, it is a product of superior performance sold at a high price in a category that no one believed existed. Friel personally attended focus groups to find out what was required to convince people that his invention would make their knives "sharp like new" and that anyone could use it. One of the things he learned was that third-party endorsement was a key. So, much of the Chef's Choice promotion effort went to stimulating newspaper and magazine articles by famous culinary arts writers such as Pierre Franey.

Cal Hodock agrees with my belief that you just can't delegate the function of observing focus groups to uninspired hacks. In addition to picking a talented, insightful group leader, he says that the sessions should be observed by the inventor—who envisioned the product in the first place—and by your own best marketing-brain person. It's not just what the round table participants are saying, it's who's listening and observing.

I asked Cal how you choose a panel for the best results. He explained:

> With a small group, that's critical. It isn't enough just to round up some people who appear to have the income or the family status or whatever to make them prospects for your product. You especially need panelists with imagination because you're not just counting noses, you're trying to get their emotional reactions, their excitement or lack of it, their criticisms, their suggestions for improvements.
>
> General Foods developed a series of questions to use in screening people to find out if they thought creatively. It didn't seem to work.
>
> I keep going back to the coat hanger test to see if people are inclined to think creatively—you know, how many products can you create out of this wire coat hanger? If they can't do at least three right away, they're not very useful panelists. In a word, it's better to screen people's behavior than to query their attitudes.

Over the years, many an inventor and many a marketing man has said to me, "Gee, if I could only get this product into a real store where real people could buy it, I know we'd have a winner." Pseudo stores where people can make shopping choices after viewing commercials or other stimuli are a part of several market research firms' repertoire.

Perhaps new-new product marketers and inventors will find an answer in Irving, Texas, where Zane Causey opened a store to sell *only* newly invented products. The name: New Product Showcase. It's an enchanted place where inventors and entrepreneurs can have any new product offered at retail, provided they pay a stocking fee. In addition, for $1,200, Causey will conduct a study of consumer response to the product, the package, and so on.

By the time you read this, Causey may have half a dozen such stores across the nation or he may be out of business. What is certain about his effort is that he has touched upon a deeply felt need, one that other bright minds should address: how to get a realistic reading on new-new products.

Uncommon Sense and Business Judgment

There is something for the entrepreneur/inventor to learn about market research from the big crashing failures of large companies with enough money to research the living daylights out of their new products.

Take the case of Premier® cigarettes. Launched by RJR Nabisco, Inc., as a "smokeless" solution to the world's annoyance with smoking, the product was based on a new technology that was said to supply smoking satisfaction, including nicotine, without any puffs of smoke clouds in the air. You'd bet a few bucks, wouldn't you, that this product had been subjected to every kind of consumer research known to the marketing world? Then why did it fail so miserably?

Robert McMath, president of The Marketing Showplace in Canandaigua, New York, who has seen thousands of new products come and go, offered me this analysis:

> They missed at least four fundamental facts. First, the actual smoke as it's exhaled in a billowing stream, wreathing around the head, is one of the major satisfactions of smoking. Premier took this away.
>
> Second, smokers like to have something to do with their hands, like flicking the ashes as the cigarette burns down. You can't do much with a cigarette that just glows on the end.
>
> Third, the Premier cigarette had an unpleasant smell for some people. It had a lackluster taste and sometimes it got too hot for comfort.
>
> Fourth, RJR hit unforeseen government trouble. Surgeon General Everett Koop declared that the smokeless cigarette was really a delivery system for nicotine and thus should be regulated by the FDA rather than the Bureau of Alcohol, Tobacco and Firearms. Health guardian organizations chimed in, and the clamor was on to require RJR to file for FDA approval for Premier cigarettes.
>
> Now when Premier was launched in the fall of 1988, those first three objections were widely voiced by consumers and dealers alike. My questions are, who were those people on the Premier consumer research panels that RJR ran before the launch? Who read the research and reported the conclusions? And who read the reports and made the decision to go to market?

McMath's skepticism about the market research is probably justified. But what about common sense? Was there a pent-up demand for this invention? Was this an invention with instant impact, one that made the first hundred people who saw it stand up and cheer?

In their book *Barbarians at the Gate* (Harper & Row, 1990), Bryan Burrough and John Helyar suggest this explanation: Ross Johnson, CEO of RJR Nabisco, was totally focused on improving the price of the company's stock. Having explored without success various mergers and dozens of schemes for pumping up the stock, Johnson opted for sending Premier cigarettes to market and letting the consumer decide its fate. Burrough and Helyar report that consumer testing had revealed major product problems, with less than 5 percent of smokers liking the taste and many people finding it downright awful.

As a lawyer friend of mine says, "*res ipsa loquitur*," which, freely translated, means, if you want to fly in the face of the facts, you'd better have deep pockets. The total cost to RJR of the Premier fiasco was estimated in news stories at $350 million.

For the inventor/entrepreneur or for the new products person in a corporate environment, the RJR tale, while an extreme example, reinforces the fundamental rule of market research. Get out there and hear and see for yourself how people react to your new product and then take note of the raspberries as well as the cheers.

Watch the Blind Side

Another incredible example of how major market research, in the absence of finger-tip business "capeesh," can mislead corporate management was the launch of New Coke®. In the process of deciding if it should replace traditional Coke with its new formula as a means of fighting Pepsi-Cola®, the Coca-Cola Co. held taste tests involving more than 200,000 people. That ought to give you statistical reliability, shouldn't it?

It should, but there was a problem. The objective of the new product was to take business away from Pepsi. The research people were so focused on whether people liked the taste of the new formula in blind comparison tests that they forgot to find out how many people were hooked on the taste of good old Coke. They forgot the fierce loyalty factor. Most of the tastings were conducted with large groups, and the participants were never told, "We're going to eliminate the regular Coke you've been drinking when we introduce this new formula."

In the long run, however, the Coca-Cola Co. executed a coup—

rather like a chess game—by reintroducing "Classic Coke" and thus occupying additional slots on supermarket shelves and in vending machines.

For the inventor/entrepreneur, rare indeed is a technology-based new product found in a massive market like colas. So large-scale consumer research is seldom the right approach, even if it were affordable. For any kind of market research, however, the Coca-Cola case leaves behind a huge warning: Don't just count noses, don't be afraid to probe for possible problems, and find out what's on people's minds beyond the questions you ask.

For new-new products, there is no substitute for unfiltered feedback: observation of how people actually react to the product through firsthand, up-close, action-oriented investigation.

The Japanese, according to many sources, including Cal Hodock, outdo us in market research as much as they do in design and manufacture. Senior executives, not temporary help, get out there to watch people as they buy or reject products in stores. They go out and learn something, then go back and try to improve the product based on that bit of information. And then they repeat the process. Much Japanese electronic research is conducted by making a small run of a new product and then getting it into a few leading stores where company executives watch consumer behavior and then ask consumers how they feel about the product.

The Japanese call this process product churning. They put out all the products they think might sell and let the marketplace dictate the winners. They match or "cover" each new product a competitor launches. Of the one hundred or so new soft drinks that appear somewhere in Japan each year, perhaps ten survive. This pell-mell race to market leads to the development of Model II even before Model I hits the market.

Obviously, the Japanese distribution chain is also different from ours, and it would be difficult to emulate their product churning system. But even as we learned from their "just in time" control of components' arrival for manufacturing, we could benefit from adapting their in-market product test system to our business realities. Certainly, for the individual inventor or small company, getting the innovation to market, *and being prepared to change or modify it*, is an article of faith in my book.

Perhaps no one has summed up the danger of relying upon plain old questionnaire-type research better than Dr. Lewis Cohen. Cohen's Law states: "If you ask a question, you'll get an answer." Especially for new-new products, you'll seldom find useful truth in cut-and-dried answers, no matter how large your sample is.

Chapter 9

New Product Marketing Insights—Part II

Throughout this book, you'll notice that I espouse a marketing principle that could be summarized by Nike's neat slogan for its athletic footware, "Just do it."

What I'm saying is, the more revolutionary your product is, the less chance you have of getting accurate information by asking people what they think about it. To find out if you have a winner, you have to interview buyers, and to get buyers you have to find some method of distribution, even if you have to go out and sell the invention yourself, door-to-door.

Now there's another side to this rule, a side where premarketing research can be immensely valuable, a side that inventors very often hate. This is research deliberately aimed at finding out what is wrong with the product/technology. Just as you need stability testing, durability testing, or performance testing—whatever technical assessments are possible—you need to test ultimate buyer criticism. It will save heartbreak and money in the long run.

Bob McMath gave me a case in point, one that proves that even the Japanese have marketing blind spots.

The product was a Nipponese invention, a single-use paint container-with-applicator. It was a small box, about the size of a single-serve cereal box with a sponge attached to one end. To paint with it, you broke the seal and, as the paint oozed through, you just wiped the sponge over the surface. Most people on the consumer panels thought it sounded great. No more spills, no separate brush to buy and clean, not too expensive. It was a winner—until someone asked the last question on the list: "What would you paint with this? What jobs do you think you'd use this for?"

The whole idea fell apart. People began to think, "Gee, it's too small to paint even one wall, so I'd need several, and then it gets expensive. Also, how do you use this thing in corners? If I don't have something big enough to use it up, I can't save it for the next day." And so forth.

As my astute wife keeps reminding me in business deals, always look for the problems revealed by research; the good news will take care of itself.

Finding the Channel to Market

At the core of every successful invention introduction is a perfect mating of the product itself and how people perceive it. Getting the world to understand your product and accept it isn't always a function of a large advertising budget. Here are a few examples of creative solutions to launching new products.

Singer's Electric Sewing Machine

When Singer introduced electric sewing machines many years ago, few women had ever seen such a device, let alone used one. Singer tried selling in department stores, selling door-to-door, and even selling by mail. Eventually it became clear that women were afraid to try the machines lest they fail.

So Singer created sewing centers where women could sit down and learn to use a sewing machine under kindly tutelage and where they could actually make something before they were asked to buy. In retrospect, this seems an obvious answer, but it wasn't obvious before Singer did it and it created a preeminent position for the brand that lasted almost forty years. Vijay Kothare, who helped edit this book, wrote in the margin of this manuscript page: "Singer even had these classes in Bombay in the '50s! I remember my sister's group going to one." And that's how names become international trademarks.

As time went on, of course, a generation of women grew up who learned how to use a sewing machine from their mothers or in school. Then the market was open to selling machines through department stores, Sears, shopping malls, and other outlets, which left Singer more vulnerable to competition. And in recent years, low-cost ready-to-wear clothes, rapidly changing fashions, and a vast increase in the number of working women have all taken their toll on the sewing machine market.

Still, Singer's principle of letting the customer "test drive" the product is a fundamental of marketing that should not be overlooked for many new products—even though inventing an economic way to do it may challenge one's imagination.

Lever's Gel Toothpaste

Sometimes an invention languishes for lack of a clear raison d'être. Such an invention may be different, perhaps even unprecedented, but people just don't visualize it doing anything they really want or need.

Toiling in the vineyards of the Lever Bros. research department some years ago, a man named Sol Gershon discovered how to make a toothpaste in gel form. It was translucent, even when the customary polishing abrasives were incorporated in it. Such abrasives would normally have turned the gel cloudy and unattractive. But Gershon discovered that if he matched the index of refraction of the abrasive with that of the gel, the abrasives were effectively made invisible.

It was a discovery. It was new. It was patentable. It had only one serious problem: Nobody wanted it. It didn't look like toothpaste, it didn't have any cleaning advantages over regular toothpaste, it wasn't better for preventing cavities. It just sat there and looked beautiful.

So the discovery went into the Lever Bros. files.

Five years later, someone at Lever had a brilliant marketing insight and told the R&D department, "Look, we've discovered that as long as a dentifrice gets the teeth clean, most people choose their brand based on how it freshens their mouth and their breath—the mouthwash effect. Could you make a toothpaste for us that would be like a mouthwash combined with a dentifrice?"

Morton Pader, the lucky fellow in the R&D department on whom this challenge had been thrust, after a fruitless period of lab bench experimentation, went to his old friend Sol Gershon to chew on the problem. Sol said, "Why don't you go dig out the file on my dentifrice gel? We even got a patent on it. The fact that it's a clear gel should make the mouthwash claim believable—it almost looks like Jell-O with Lavoris in it."

And that's how Close-Up® Toothpaste was born. It's one of many stories that remind us all that great inventions need great marketing insights to bring them to life. The problem is to get to the insight and the invention concurrently.

The Razor Blade Wars

Of course, the root marketing question for every new invention is how to select the most feasible marketing channels, that is, not just the normal avenues of distribution but the channels that are available and affordable for an upstart product.

In the early 1960s when the Wilkinson Sword people in England wanted to introduce the first double-edge stainless steel razor blades in the United States, they faced a formidable foe in Gillette.

In the first place, Gillette had an enormous investment in and commitment to its Blue Blade, a product that, until then, gave the best shave for the money. Gillette appeared impregnable. It had files of market research proving that men were very satisfied with their Blue Blades, and no Gillette executive was walking around Boston to see how many men had nicks or cuts or complaints about rough skin.

In addition to the Blue Blade's entrenched position with consumers, Gillette obviously could fight Wilkinson with its extensive sales force and strong advertising. Equally difficult was the fact that druggists, then the primary distribution channel for razor blades, had for years heard pitches from new-blades-on-the-block claiming that they shaved closer or stayed sharp for ten shaves or didn't nick the skin or what have you.

Rebuffed by major drug chains and wholesalers, but firm in their knowledge that the Wilkinson Sword Stainless Blade® would generate a huge market if enough men could be induced to try them even once, Wilkinson's U.S. salespeople began to place the unit in—would you believe?—garden supply shops. They developed a counter merchandiser unit that held a small inventory and yelled a loud message, "New Stainless Steel Blades from the makers of famous Wilkinson Shears and Clippers!" They had two things going for them: The name Wilkinson was already known to the gardeners of America for its high-quality implements; most important, the garden supply shop owner understood that the profit he was making on these blades was a dollar he'd never see otherwise.

There were unexpected benefits from this unusual distribution tactic. Garden shop owners found their customers coming back more frequently than usual, so they loved Wilkinson blades and "talked them up." Because the Wilkinson Stainless Steel Blade did indeed give a better shave and lasted longer than the Blue Blade, one man told another about his blade discovery. Meanwhile, Wilkinson, having viewed this project as merely a last-chance means of seeding the market, grossly underestimated production, leaving many stores out of stock—which, of course, only served to increase demand and generate more consumer word-of-mouth advertising.

Within a year, drugstores began to stock Wilkinson Blades and Wilkinson established a place in the American market. Within two years, Gillette announced its great innovation: stainless steel blades.

To Gillette's everlasting credit, its management decided that blade innovation cuts two ways. The introduction of twin-bladed Trac II® razors in 1971 and swivel-headed Atra® razors in 1977 increased Gillette's share of the market. The appeal was not simply that two blades are better than one; these razors shaved better, nicked less. Unfortunately, Gillette could not stop Schick, Bic, Wilkinson, and others from making double-bladed razors.

Then Bic, with its automated manufacturing skills, pushed hard on disposable razors, forcing Gillette to make its Good News® disposables a high priority. As luck would have it, disposables, the lowest-margin product, became the real battleground, chewed up large chunks of the marketing budget, and drained Gillette's profit margins.

Then, in early 1990, having invested ten years and $200 million in research and development, Gillette introduced the Sensor® Shaving System, using a razor with two blades suspended on tiny springs that permit each blade to float independently as it follows the topography of the skin. Armed with seventeen patents, Gillette priced the Sensor to provide hefty marketing margins, captured 9 percent of the blade market, and made the razor profitable in its first year. Furthermore, this unique product enabled Gillette to improve its position worldwide, even to gain market share in Japan, the only overseas market where Gillette trailed its longtime rival, Warner-Lambert's Schick blades.

No doubt, Schick and others will find products to compete with Sensor. But in an era when so many people, including this author, have bemoaned the corporate constipation that has cost American companies their technological leadership in so many fields, it is important to salute a shining example of a major company that had the creative talent and the management commitment to do it all right.

Another reason for recounting these snippets from the Great Razor Blade Wars is that razors and blades are historically a favorite turf for inventors. Men, long suffering from scraping their follicles, have for years imagined extraordinary ways of removing their facial hair. But few have made money from such inventions in recent years—unless you count what Victor Kiam did to improve the Remington Electric Shaver and rejuvenate the company through his extraordinary personality on TV. It would take an invention with dramatic performance advantages, proved producibility, and a clear patent position even to secure a meeting, much less a license agreement, with a Schick, a Wilkinson, or a Gillette. And the monster capital investment required for start-up makes the razor business an unlikely field for the inventor/entrepreneur.

Quite possibly, by the year 2000 someone will have come along to prove me wrong, not because they have discovered a fabulous new razor but because they have found an entirely different technique—mechanical, chemical, or electrical—for removing or stopping the growth of facial hair. Short of such a breakthrough, however, improving the lot of unbearded males will continue to be a hobby and not an attractive business.

Using Hindsight for Foresight

Similar criteria apply to other fields. When you consider the commercial potential of an invention—before investing time and money—it seldom pays the entrepreneur/inventor to put a big effort into fields that require heavy capital investment and long R&D gestation periods, for example, paper towels, automobiles, copper tubing, to name a few.

The Great Razor Blade Wars also offer valuable insights for corporate managers, the most significant of which is this: A new product based on technology that delivers discernible benefits to the user is the marketing man's best friend.

When you have a superior product, like the Wilkinson Stainless Steel Blade, find some way to get people to try it. If you can afford to sample it, give it away like popcorn (see the story of Smartfood later in this chapter). If it costs too much to give away, look for a unique distribution method that will build a core of enthusiasts; financier Al Bianchi "sampled" Lotus cars in the United States by giving one to every investor in the U.S. distribution company who put up approximately $250,000, giving them an apparent 20 percent discount on their investment.

If you're a company executive living with pressure for performance, it's interesting to note that Gillette was the target of several takeover attempts while the Sensor was in development. To fight off a not-so-tender offer from Ron Perelman's McAndrews and Forbes, Gillette executives used all the devices they knew, called in all the favors they could summon. Yet they knew that, in the long run, it was only a successful, dynamic, profitable Gillette that would boost the price of the stock and keep the loyalty of stockholders. So, while the predators were at the gates, they poured millions into the development of Sensor and built a takeover barrier based upon profitable, protectable technology.

One final thought about the blade stories: Suppose you were an independent inventor or a small development company and you had the brilliance to envision the floating blade razor. It's not likely that you could raise the resources necessary to complete and prove the invention, including high-speed manufacturing, or to underwrite the high costs of market entry.

Is there a way to make money from such an invention? Here's what I'd do. Once I had patents well thought out and applied for, I'd go to Wilkinson in England, or Bic in France, and license the invention to them, royalty free, for the entire world excepting North America, in return for their developing and producing the product—and agreeing to supply me with all the razors I could sell in the United States and Canada.

Differentiating New Products

One thing that Wilkinson Blades, Close-Up Toothpaste, the Interplak Electric Toothbrush, and the Singer Sewing Machine had in common was that each was based upon a genuine invention, something physically different that the user could perceive as functionally different. With imaginative strategies, such products can succeed against both tough competitors and market inertia.

On the other hand, products based upon extremely minor advantages or resting upon product positioning alone need bigger introductory investments and are more likely to fail. Such marginal innovations are frequently confused with genuine inventions by marketing and manufacturing people alike and are a prime source of red ink.

Visualizing a Target Market

Some years ago at Hood Dairy in Boston I overheard a conversation that went something like this:

Product development man Vanilla ice cream still sells best, but we ought to do something new with it. A guy came in the other day and offered us lichee nuts at a terrific price, so why don't we put lichee nuts in vanilla ice cream? That's never been done before.

Agency account executive Say, that's a great idea and I've got just the name for it—Chinese Vanilla!

What's wrong with this scenario? Pretty obvious. The proposed product is still "plain vanilla" and no target market comes to mind, excepting perhaps Chinese restaurants.

Compare this with what a small company in Brooklyn did with a product called Tofutti®, a frozen product made of tofu, high in protein and low in cholesterol, and derived from beans. Most important, Joe Eisenberg, the man who started making Tofutti in his kitchen, found a secret way to make this frozen bean stuff taste delicious, something that others had found an impossible task. Joe didn't do any formal market research, but he did have a clear vision of his target market: people who wanted a frozen dessert with more substance than Italian water ice and fewer calories than Haagen-Daz ice cream and who would be willing to pay for it. Tofutti, cleverly introduced at upscale places like Bloomingdale's Forty Carats restaurant, a "healthy food" spot, took New York by storm.

Not to worry that Tofutti's secret formula was matched by competitors within a year of its blast-off; not to grieve that other manufacturers found ways to make frozen yogurt taste as good or better than Tofutti or that people discovered that Tofutti was really not all that low in calories. In 1984, Tofutti, Inc., went public for the sum of $18 million, proving that tofu, after all, is more than just beans and providing capital for other new product launches such as Tofutti Egg Watchers.

Inventing the Package, Packaging the Invention

Just as sometimes the medium is the message, there are times when a packaging invention makes the product "new." Occasionally, such packages make the inventor rich and even famous.

The familiar pop-open can with the attached pull-ring—the very one you've opened a zillion times for your soft drink or beer—was invented in 1960 by Ermal Cleon (E.C.) Frase, CEO of the Dayton Reliable Tool and Manufacturing Company. E.C. was out on a picnic without a can opener and found himself opening a can of beer on a car bumper. It bugged him. A few nights later, too much good coffee at dinner kept him awake and he sat down to doodle ways to solve the can/ring problem. By morning, he had the answer.

Pull-ring openers that came apart from the can were well known. But sloppy people discarded them all over the landscape and communities banned products using such menacing bits of sharp metal. What Frase conceived was a way to rivet the ring to the can end. Most important, he knew how to do this on high-speed can-manufacturing machinery at a minimum cost.

The invention was, like many great ideas, essentially simple. He visualized how you could make a small nib or rivet on the can top by re-forming the same flat metal that makes the can top itself. Then it was an easy next step to attach the pull-ring to that rivet.

An amazing sidelight is that Ernie Frase—who had never gone to college but was graduated from the General Motors Institute with trade skills—created this invention in his mind with a mathematical formula to calculate the formability of the aluminum.

Now how do you market an invention like this? Frase was more fortunate than most new package inventors. Alcoa was in the midst of a program to develop new markets for aluminum and, back in 1960, steel cans had a firm grip on the beverage industry. When Ernie Frase made a cold call on Alcoa's engineering department, he simply asked them for several different aluminum alloys so he could use them to test his pull-ring idea.

Not unexpectedly, the Alcoa engineers told Ernie that it wouldn't work. In fact, when three Alcoa men first saw the samples, one of them said to the others in a stage whisper, "I think we're being hoodwinked."

That initial distrust, however, turned out to be a blessing because, when Ernie was able to produce functioning samples with Alcoa's own tooling—tooling that he had corrected himself—the word went zipping to Alcoa management, which promptly moved to tie up the invention. Thereafter, Alcoa engineers teamed up with Frase to solve the practical problems of converting the invention to a high-speed manufacturing system.

The business deal was a doozie. Ernie Frase was primarily interested in building specialized manufacturing equipment and acting as a supplier of lids. So Dayton Reliable Tool and Manufacturing Co. got that end of the deal and it made the company and Ernie Frase a power in the specialty machinery world. In addition, Alcoa licensed Frase's patents and then sublicensed almost every major can maker in the United States and around the world.

The royalty on the patent license was only a fraction of a cent per can lid, but with about 150 billion can lids being produced each year, Ernie learned to live with that. In fact, he could live just about any way he wanted to, because the royalty payments went not to the company but to E.C. Frase personally. But he was a modest man and lived that way until he died in 1989.

In addition to eliminating the litter of loose pull-rings, Ernie Frase's invention did some good things for the economy. Hank

Bachmann, a longtime associate of Ernie Frase, told me that this invention is responsible for about 500,000 jobs around the world and a plant investment of some $7 billion. Another nice thing: 55 percent of Dayton Reliable's business is overseas, bringing a steady stream of income back to the U.S.A.

Another inventor whose ingenious packaging has almost surely passed through your hands is Sanford Redmond, the man who created the little butter pat trays with the wax paper on top, the ones you find next to your bread in so many restaurants. Most creameries today use his machines to make those butter pats. That invention built his first family fortune.

But the invention most likely to endear him to millions of people is called the dispenSRpak. It's the first portion package that you can open and dispense with one hand—an individual serving package with applications for food, cosmetics, drugs, and many other fields worldwide. Redmond might have settled for selling to institutional food suppliers or fast-food chains, but he looked for major manufacturers for which his dispenSRpaks could create genuinely new product opportunities. Just as going after the easiest-to-sell application is right for a new business, using your imagination to look at *all* the possibilities before hitting the most obvious one is good invention marketing.

The story of Sanford Redmond's start-up is told in some detail in Chapter 13.

Despite such spectacular successes as Frase's and Redmond's, it is often difficult for the independent inventor to sell packaging inventions. There is a Catch-22.

When you create a packaging innovation and go to a marketer with it, even if he says, "Hey, that's interesting!" he will also say, "But we don't make packages. If some supplier offers it, we might be interested."

Now you go to a package manufacturer and say, "This marketer really likes my invention—how about if you people take a license from me and engineer it and manufacture it and we'll both make a bucket of money?"

"Sure," says the package maker, "can you get me a firm order for the first x-million units?" Back to square one.

Of course, that's not universally the case. Some folks have bridged the gap between package manufacturer and marketer by targeting an acute need. Rich DiCicco of Technology Catalysts told me, for instance, that an independent inventor, Art Harris, developed a very accurate metering valve for pharmaceutical aerosols and sold it to the one company that could best use it, American Home Products, for

Primatene Mist, an asthma product that demanded accurate dispensing for safety reasons.

But Frase with a Big Brother partner and Redmond with the ability to make and sell the packaging equipment are examples of two surefooted ways to wealth in the package-invention world.

The Myth of New Product Failure

It's important for inventors to remember that many marketing and advertising people together with the media are notoriously sloppy about what they term "new products," and thus the reported rate of failure for new products is sky-high. Time and again, *Business Week*, *Advertising Age*, *The Wall Street Journal*, or equally prestigious publications bemoan the high rate of new product failures and berate the executives involved. The headline of a full-page UNISYS advertisement in the November 2, 1988, *Wall Street Journal* read: "80% of new consumer products won't survive one year in the marketplace."

Where do those shocking statistics come from? Usually, the story quotes somebody's survey or somebody's opinion, although the UNISYS ad didn't quote any source. Recently I went through back issues of *Gorman's New Product News*, a bible of "new" package goods introductions, to check the raw facts on which such frightening failure statements could be based. What I found was this, that the large majority of these "new products" were new flavors of existing products, new sizes of existing brands, new packages for existing brands, new brands that merely copied products already marketed by someone else, and, I suspect, in some cases just new "positionings" for old products where the only thing new was the word *new* on the package. (No wonder retailers resist the flood of such new products and impose slotting allowances.)

While I recognize the dilemma of a marketer who cannot find a new word for "new," the marketing costs for launching copycat products are frequently disproportionate to the payout. Time and again, buyers reject old products in new guises or disguises. To report the failure of me-too products as the norm for all innovations does a disservice to invention in America—and certainly makes it tougher to finance inventions.

Invention Marketing 101

Because I have helped launch new products from sausages to antacids—so to speak—people sometimes say to me, "There are so many different ways in which innovators have marketed new products. Aren't there any *basic* rules?" Or, as my redundant but pungent former client, Les Rosskam, once put it to me, "Yes, but what is the nub of the crux?"

You might expect that I'd equivocate about providing a simple answer to a question on which experts have written volumes. Not so. There is one golden rule that rises above all others in the successful commercialization of new-new products, products with performance differences based upon invention.

That rule is: *Get the product out of the testing and development stage and into the marketplace as soon as you can.*

Sounds simple, but it isn't. Implied in this principle is that you will find a way to get the product manufactured, even if they tell you that the tools and dies don't exist, or that you cannot meet minimum quantity requirements, or this or that. You'll do whatever product testing and customer research is helpful—but only for the purpose of improving the product before the actual test in the marketplace, not to decide the product's fate. You'll substitute ideas for money.

Still, those are only the necessary preliminaries. The moment of absolute truth comes when people lay down their hard-earned cash to buy the final product—and then express their delight or dismay. When genuine purchases take place, financial pundits, marketing wizards, retail buyers, skeptical suppliers, and even seasoned reporters are impressed. So find a way to market, or create one.

If you can't get the product into the normal channels of distribution, you'll sell it through unconventional routes until it has proved itself. The way products are sold changed radically in the 1980s and it is only the beginning. Many department stores have lost their niches. The big drug chains have become specialty department stores. "Membership club" warehouses and manufacturers' own stores in outlet malls have moved customers away from their habitual shopping places.

On top of the changing patterns of store retailing, telemarketing has proved effective for many consumer and industrial products—even though maddening to many people at dinner. Franchising proprietary systems from accounting to auto engine tuning is an alternative route. And direct marketing has been refined to a science both through computerization of mailing lists and sophistication of mailed presentations.

All these marketing channels are important in themselves; they can also lay the groundwork for follow-up sales efforts through conventional retail channels.

Catalogs with specialized points of view have been able to sell everything from pet foods to $3,000 wine cellars, and sometimes create product recognition far beyond the actual orders received. Bear in mind that a catalog like Hammacher-Schlemmer's goes out to over 3 million people, most of whom love to read it whether they buy or not. Dan Friel told me that the appearance of his Chef's Choice knife sharpener in many catalogs was his most effective advertising because it not only reached millions of people but gave the product a stamp of acceptance.

Home shopping via television and so-called information-advertising TV programs and direct selling via computer networks have all added new distribution opportunities. With home VCRs now commonplace and videotapes so inexpensive, mailings of videotapes have become a viable medium for the direct selling of appropriate products. And because mailing lists can be selected by zip code to match almost any demographics, pinpoint testing of new-new products can be done with relatively low investments.

Whether you launch on a large scale or small, once you get your new product out there into the hands of real purchasers, you can find out if it deserves to live or die—and why. If you really love your innovation, you won't hesitate to get out there and learn what's happening. Is it moving at the rate you expected and, if not, why not? Do the people who bought it love it, hate it, or put it aside and forget it? That's the time to spend the market research time and money.

To Market, To Market

I know of no story that better illustrates this point than what happened to Post-it® Notes in test market.

Art Fry, a senior new product development man at the 3M Company and the godfather of Post-it Notes, told me that 3M's market research people predicted to management that the total market for this product would be between a half and three-quarters of a million dollars a year. Pragmatist Art Fry had trouble believing this research. "If the market research was right," he said, "it would hardly pay 3M to go into the business."

In Fry's R&D laboratory, the Post-it Notes were being distributed to one and all to find out how people would use them. Art checked the amount being used and found that lab people, strictly the ones not connected with his project, were using more Post-it Notes than Scotch® Tape, as measured by dollar value. And Scotch Tape was the largest-selling single product in the company! Armed with this information,

Art Fry and his team VP, Joe Ramey, convinced 3M's management to go to test market.

Everything went according to plan: beautiful point of sale material, strong local advertising, and excellent distribution. And, in Art Fry's words, "It went over like a lead zeppelin."

Marketing people from 3M had visited Richmond, Virginia, a prime test area, and reported that dealers all said, "It's a nice product, but it just isn't moving." Management was ready to pull Post-it Notes off the market when Art Fry and Joe Ramey asked for time to go out in the field and find out why the product was bombing.

They did something every marketing person should remember. Instead of just talking with the trade, they called on some companies that had actually bought the product and found out what it was being used for and by whom. It turned out that almost everybody had a different use for the little tack-up papers. The product brought out the creativity in each individual, and they felt very good about using Post-it Notes that way.

Without waiting to return to 3M headquarters to report these findings, Joe Ramey organized a sampling program to send Post-it Notes free to a number of leading Richmond companies. Within a few weeks, orders from corporations began flowing back to the stores, a trickle that has turned into a mighty stream, making Post-it Notes one of the five largest-selling items in the office supply business.

Nor is this principle lost upon smaller businessmen. When Ken Meyers and Andrew Martin launched their cheese-flavored popcorn called Smartfood, they realized they could not afford enough advertising to affect the market. So they hired a marketing agency, Target Marketing & Research, Inc., to place people dressed in Smartfood T-shirts, boxer shorts, and sunglasses to sample the popcorn outside 7-11 stores and later in large supermarkets. People could eat the stuff and, if they liked it, buy a bag on the spot. A Smartfood van was parked outside to resupply the store as the product moved out. From May through Labor Day, 500,000 bags of Smartfood had been sampled away—and four years after the company was started, Meyers and Martin sold it to Frito-Lay for $15 million.

To repeat and amend the golden rule for Invention Marketing 101: Get the product out to market, any way you can—and then follow it as if it were your own child crossing the street.

Merging the Product and the Concept

If you've ever had your eyes examined for glasses, you may remember a test in which you see two images and the doctor moves the lenses and asks you to say when they merge into one.

That's a fair metaphor for a product invention and a product concept. The former is a tangible thing, the latter a superimposed vision of how the thing performs, feels, satisfies. In order to sell new products, especially new-new products, it's important that the two images, the invention and the concept, align perfectly.

Truly great introductory advertising and major public relations coups do more than get attention. They project the gestalt of what this new product is and does. One example: that very original adman Hal Rainey's campaign introducing GM's new Saturn automobiles—a car that you don't see but rather feel through the words and pictures of real-life people who are building it and who care a lot about how it comes out. And that attitude was underscored by GM when, faced with glitches in some of the early production, the company replaced the cars rather than just recalling them.

In the early 1960s I ran an advertising agency and had an opportunity to devise an invention and its image simultaneously—a truly integrated product and personality. Our client, Abbotts Dairies, a division of Fairmont Foods, asked us to find a way to sell more milk and to sell it more profitably, no small challenge.

James Russell, the agency's account executive for Abbotts, and I did an in-depth review of all the possible market segments for milk. Several years before, I had devised a cottage cheese diet plan that was a surefire rocket for that product every spring, just before bathing suit season. Now we were searching for other specific areas where a new use or appeal could boost sales.

One of the Abbotts executives asked us what we could do for its fortified milk. It had a number of vitamins in addition to the natural vitamin D of milk. But the product was a turnoff. It was sold in a brown bottle, needed to protect the vitamins. It came only in half-gallons. And it retailed for three cents more than regular homogenized vitamin D milk. It had all the earmarks of a product for the weak and the sickly.

Russell and I wrestled with the problem until one of us asked the right question, "Who drinks most of the milk, anyway?" Answer: children. Product answer: Create a milk that is *better for them* than regular milk.

We went to the Abbotts laboratory with a proposal to put into a single quart of milk the recommended daily children's requirement for all vitamins—so that "if your child drinks a quart of Abbotts milk, there is no need to worry about vitamin pills." *And* make sure that there is no change in the taste of the milk.

While the laboratory people were formulating and testing the product, the agency people went to work on a name and a package. "VitaMilk" wasn't available, but we came up with the name Mighty Milk and the art department designed a colorful, fun milk carton with side panels to be used for educational kid games and public service messages. The brown bottle disappeared.

Then we hit a snag. Who would pay for the cost of the vitamin package? If we passed the cost and a markup on to the consumer, research showed, it would severely limit sales. If Abbotts took less profit per package, Mighty Milk would have to increase its volume 25 percent to stay even on net income.

The answer was obvious—once we thought of it. At that time, low-fat milks were just coming onto the market. People were just starting to say, maybe we don't need all that fat. We checked with nutritionists, who told us that children would be very well served with half the 3 percent butterfat in standard milk. Fortunately, there was a market for the 1½ percent fat Abbotts took out of the product—it was used for butter—and the net cost of Mighty Milk actually dropped a half cent a quart below that of regular milk. That provided an increased promotion budget and, after the introduction, a blip in profits.

Mighty Milk is still being sold some thirty years later. It shows that, even with a commodity like milk, an invention that offers a genuine benefit combined with a concept that is right-on for the market can satisfy everyone—except one's competitors.

Innovation Thought-Starters and -Stoppers

When you examine all the examples in this book and hundreds of others, it's very tempting to look for a new product success formula. But it doesn't exist. What does exist are people with great inventive minds and people with great business imagination; seldom are both found in the same body. Such people invent their own success stories.

There are, however, two fundamental steps that can't be skipped in launching invention-based new products, and they have to be taken in sequence:

1. Dig in on your hands and knees to determine if you really have a winner—and not just a competitor—and do this early in the game. Make sure your invented product has instant impact. Do the kind of market research that is useful for genuinely new products, not the kind used for variations on old products.

2. Once you're satisfied that you have the product right, that people who see it will reach for their wallets or checkbooks, find the unique marketing strategy that suits your product and operating budget, a strategy that will make third parties want to get behind your product and push, and one that doesn't unduly alarm major competitors before you get a foothold.
The shorthand for this is having a *leveraged marketing strategy.*

It's that second part that gets sticky, but the people who build fortunes on inventions almost always go off the beaten path in some respect to get their creation to market, despite a lack of funds or organization.

A classic example is Loctite Corporation, a company doing over $400 million in sales in 1988 by selling more than a thousand chemical specialty products in some eighty countries around the world. The company was started in the early 1950s; the initial capital was $100,000 of convertible debentures; first-year sales were just $7,000.

Just in case you have never heard of Loctite, here are a few details. Vernon Kreible was in his final years as head of the Department of Chemistry, Trinity College. When his son Bob, who worked for General Electric, told him of a synthetic monomer, tetraethylene glycol dimethacrylate, that would stay liquid in a bottle only when constantly aerated, Vernon Kreible was fascinated. Seeing no practical application, GE sold off the monomer. But Vernon, working in his laboratories at Trinity, found that when you applied the liquid to nuts and bolts, it would harden and hold them together better than a lock washer. He pursued the idea of splitting the chemical into two components that could be combined just prior to use. Then, son Bob pointed out that air-permeable polyethylene bottles would provide the necessary aeration, and experiments proved that the material could indeed be marketed as a one-component sealant. The discovery led to the first of what is now a portfolio of more than 850 patents.

For readers who are curious as to how a liquid could hold two metal parts together, Kenneth W. Butterworth, chairman and CEO of Loctite, provides a charming metaphorical explanation:

> Raise your index finger and then grab it tightly with your other hand. Now why is it so hard to pull your finger out of your fist?
>
> Well, it's friction. If you oil your hands, your finger will come out easily. Loctite works the opposite way—it increases friction.

> How? Well, no matter how tightly you hold your finger, there will always be some inner spaces because your hands and fingers have wrinkles and microscopic surface roughness—just as do all metal surfaces. Loctite liquids flow into all that roughness, then harden because there's no oxygen available. The hardened Loctite keys the two surfaces together. Your finger and fist analogy is similar to a nut and bolt, or a press fit. And it resembles shaft-mounted parts so common to rotary equipment. Most of the problems of leakage, looseness, and wear that bedevil machinery can be traced to that microscopic inner space, regardless of even the most precise machining.

As with so many inventions that defy preconceptions, Loctite's major marketing problem at the outset was credibility. Fortunately, the product had inherent drama that appeals to the press.

Loctite was literally introduced into the market by two kinds of public relations—on the one hand, the kind of PR you get from press stories, and, on the other, the kind you get from getting out in the field and working hand in glove with your customers.

A masterstroke was accomplished almost at the company's inception. In July 1956, before the first patent was issued but after it was filed, a press conference was held for twenty-five editors at the University Club in New York. The announcement that Loctite had solved the age-old problem of loose nuts and bolts intrigued writers from newspapers and the business press. The flood of publicity generated an avalanche of inquiries, many of which were quickly converted to sample orders.

In particular, one article in *Fortune* in 1956 paid dividends for years. A distributor in Canada and another in England each read the story and took on the line; and both companies grew to the point where they became major acquisitions for Loctite Corporation. And the benefits of the publicity were felt as late as 1970 when Loctite began courting the mechanics who prep cars for the Indianapolis 500. Loctite management was astounded and delighted to learn that the company didn't have to pay a penny for endorsements because these people had learned about Loctite from that *Fortune* article and had been using Loctite on their racing cars for years.

The other kind of public relations that helped Loctite was the highest form of salesmanship. By hiring representatives who understood specific industries, Loctite was able to form close working relationships with its customers' engineers and plant people; they specified new varieties of Loctite together, leaving little room for competition.

As noted, Loctite's fundamental problem was credibility. One early management decision helped create believability and profitability in one fell swoop: Rather than adopting the usual chemical specialty markups, Loctite priced the product high right from the start, making it directly competitive with lock washers. The recommendation to do this came from a study by the Battelle Memorial Institute of Columbus, which foresaw only a limited market for Loctite. But, as Kenneth W. Butterworth, chairman of Loctite Corporation, pointed out in a speech to The Newcomen Society,* "It was selling 'to value' rather than manufacturing cost that produced the funds needed to conduct the communications and educational programs that the revolutionary product required."

By definition, an invention is genuinely new. To some degree, it is out of the range of people's prior experience. It works differently. It looks different. It may require the user to change his or her habits or beliefs. Therefore, prejudging how the new product will be accepted in the marketplace, deciding how best to introduce it and how to measure its success or failure are challenging tasks. But it's worth it to the inventor/entrepreneur and to the major marketer because these new-new products, when linked with new-new concepts, have not only better odds for winning but better staying power.

*The Newcomen Society of Engineering and Technology is a repository of information on the application of capitalist ingenuity to invention.

Part IV

Pathways to Millions: The Selling of Invention

Chapter 10

Licensing: The "Easy Route" to Money

Picture yourself at the command station of a 55-foot yacht, your captain's hat tilted jauntily as you maneuver the boat alongside a marina dock, where eager hands tie her up. The dockmaster awaits you and hands you the accumulated mail. You flip through the envelopes quickly until you come to one that you rip open, extracting the one-page accounting and a check.

"Hmmm," you muse, "not too shabby. Three hundred and forty-six thousand for the third quarter; I guess we're doing all right."

The problem with this story, of course, is that it isn't you—or even me—that I'm talking about. But it is a real person and it's only a slight dramatization of the facts about Harvey Phipard. I told you some of his story from an intellectual property standpoint in Chapter 6; his adventures in licensing and defending his rights follow shortly.

Of the three ways to make money from an invention—selling it outright, granting a license, or starting a new company—licensing appears to be the easiest route. That dream of checks in the mail is alluring. But, like any other business transaction, it has its ins and outs; and this chapter is mainly about those turns in the road rather than just the basics.

When I say "license," here, I'm talking about an agreement that grants someone the right to practice your invention, or, as most contracts put it, "to make, have made, use and sell" products produced in accordance with the invention in return for a certain payment, usually a royalty on sales.

The alternative is to sell the whole invention outright for a fixed sum, and many an invention owner would like to sell out and go fishing, but in my experience this seldom happens—unless the end product is already a substantial success in the marketplace. The reason is simple: No company executive wants to pay a large chunk of money for an invention, no matter how promising, that might nevertheless bomb and leave him with egg on his face. On the other side of the fence, no invention owner wants to sell out and turn over his brainchild to a foster parent without seeing serious money on the table at signing.

So licenses, with royalties paid over time, remain the dominant means of forming an alliance between people who create technology-based new products and those who wish to manufacture and market them. And licensing has become a professional business with hard lessons to be understood and pitfalls to be avoided.

So much money can be generated through licensing that major corporations—even companies like IBM that once clutched their patents to their corporate bosom—have turned licensing into separate profit centers. At least one company, Texas Instruments, Inc., was so successful in licensing its inventions that, at one point, its royalty revenue, enforced by aggressive legal action, became its largest single source of income. This was annoying to many who were forced to take licenses but lifesaving to TI.

Naturally, any view of the licensing process is influenced by which side you're coming from: the licensor's or the licensee's.

Because the licensee is usually a company, often a large one with an abundance of lawyers and licensing experts, most of what you'll read here is from the viewpoint of the licensor/invention owner. Despite that, or perhaps because of that, the attitudes and experiences reported here may be equally stimulating to people on the licensee side of the table.

The Licensing Preliminaries

Let's start with the bottom line. The licensor is selling and the licensee is buying. In that sense, they're opponents. Yet good lawyers will tell you, the contracts that get signed and stay signed reasonably balance the interests of the parties. That's what negotiation is all about.

Almost always, long before you reach a licensing discussion, you will need to disclose some details about the technology, the product that the technology makes possible, the tests performed, the suggested marketing strategy, and so forth. No single disclosure policy suits all situations. How much to disclose and when depends upon what is really necessary to make the deal. It also depends upon whether the parties have compelling mutual interests that may deter thievery or even honest disagreement.

I've been at presentations at which the R&D people ask endless questions, as though they were quizzing a Ph.D. candidate. Prideful inventors sometimes answer all such questions—needlessly. A pre-meeting session of everyone on your team should set the ground rules for deflecting questions relating to sensitive areas.

Nor does it hurt to knock back those who challenge you. I recall with some joy a large meeting at a major corporation during which one particularly pompous engineer declared, "I have seventeen years' experience in this field and that invention simply won't work." To which my partner, Gene Whelan, sensing the hopelessness of a temperate response, replied with true Irish wit, "No. You don't have seventeen years' experience. You have one year's experience—seventeen times!"

In any event, you and your prospective licensee will want to sign a confidentiality agreement in advance of the presentation. Every company tends to have its own pet form. However, it's best to try to use language that is fairly balanced for both parties. At the end of this chapter, I include two letters of agreement for confidential disclosures: one for an invention owner making a disclosure, and the other for mutual disclosures where the invention owner and the prospective licensees are exposing confidential information to each other.

Now let's turn to some unusual real-life cases of what people invent and how they get their just desserts.

A Many-Faceted, Happy Example

I first talked with Harvey Phipard by radio-telephone while he was aboard his yacht, summering somewhere along the coast of Massachusetts. His home is in Florida, and Harvey and his boat are found in those waters in the winter. Where you do not find Harvey, ever, is working at a desk or in a factory. That's because he gets royalties on his inventions of about a million dollars a year, and he' been getting that kind of money ever since he retired some fifteen years ago.

In his former life, Harvey was in charge of production for the Continental Screw Company, a classic, stodgy old New England company that made machine screws and, in particular, sheet metal screws, which, as you may know, have helical grooves from their tips to their necks.

Because such screws are made of a harder metal than the aluminum or steel sheets they are screwed into, they cut a dovetailing groove into the metal sheet, thus holding together whatever you're seeking to attach.

Before Phipard, however, there was a frequent problem. As such screws rotate into the metal sheet, they can bite off shavings that bind the screw or break the grooves being cut in the sheet metal. One night Harvey Phipard was sitting at home at his desk doodling when an idea hit him. His pencil drew a top view of a conventional screw: It was round. Then he drew the same view with three lobes on the circle. Then he drew a perspective view: The screw still had grooves and tapered to a point, but there were three little bumps on the outer edge of the grooves from top to bottom.

Harvey figured, "If only those outside lobes contacted the sheet metal instead making constant contact all around, it would reduce friction, wouldn't chop up the sheet metal as much, and, because the groove that this design cuts is fairly deep, it would have an increased amount of sheet metal for the screw to hold onto."

Another advantage emerged when the design was tested: The effort required to drive such screws was less than what was needed with a conventional sheet metal screw.

At this point, you might say, as I did to Robert Frank, the Boston trial lawyer who represented Phipard in later litigation and who told me parts of this story: "Wait a minute now. Harvey was an employee of Continental. He was working for the company when he got the idea. How could he license that invention?" Frank replied:

> Of course, that's the question. Exactly what was his job? My understanding is that he had a pre-employment discussion with Continental in which they agreed that his job would be manufacturing, not inventing.
>
> Maybe neither party foresaw that the invention would be big money. In any case, they worked out the deal to pay him a bit on Continental's own production and to share licensing income, each party making compromises in order to get on with the matter. So Phipard patented that invention and assigned it to Continental and, over the next few years, created some follow-on patentable inventions relating to multilobular screws.

Think about this. Especially today, when America is looking for ways to regain its industrial creative leadership, the issue of who owns rights to what, as between employee and employer, is of keen interest. Please note that Japanese law protects employees by making void any assignment or exclusive license of patent rights in advance of invention—as long as the invention is not in the specific field for which the employee was hired. Germany has somewhat similar laws to protect inventors. In the United States, the decision as to who owns intellectual property rests upon common law and a plethora of court decisions, which makes it important for the informed employer and the intelligent employee to set up in advance their ground rules for invention ownership.

In the late 1960s, Harvey Phipard decided to retire. He could well afford to. The agreement with Continental would terminate only with the last-to-expire patent. He was receiving, and would receive for many years to come, checks for several hundred thousand dollars each quarter.

But nothing is ever so easy. You know that.

Harvey Phipard faced one more major legal battle before he could relax and enjoy his yacht. Continental Screw was taken over by Amco, a company started by Royal Little when he sold out Textron; and Amco in turn sold it to Amca (a Canadian-controlled company and no relation to Amco).

Soon thereafter, Harvey told me, executives at Amca had a meeting that went like this: "Who is this Phipard? We've never seen him. All we do is pay him millions. So cut him down to ten percent of what we've been paying him and see if you can't think of a reason for doing it!"

Amca took the position that they didn't have to pay Phipard any more because the first of the patents had expired and you can't extend a patent license beyond 17 years with minor improvement patents. When Amca filed for a declaratory judgment, Harvey came running to Bob Frank at Choate, Hall & Stewart in Boston. There ensued a fascinating legal joust.

Bob Frank discovered that Amca was still licensing Phipard's follow-on inventions to other screw manufacturers and collecting hefty royalties. "If they're licensing third parties," Frank reasoned, "they wouldn't stop paying Phipard on those same patents without first seeking an opinion from their counsel."

Naturally, Amca's counsel argued in court that communications between lawyers and their clients are confidential. "But this is different," said Frank. "Amca has told the court that it relied on the advice of its counsel in terminating Phipard's contract. So we are entitled to know what that opinion said."

After a large legal wrangle, the court ruled for Phipard. Among the papers turned over in the discovery proceedings was an opinion provided to Amca by its lawyers that said exactly what Frank deduced, namely, that there was no legal basis for refusing to pay the royalties. Amca was dead in the water.

There is a twist to the end of this story. As often happens in such bitter battles, events overtook legal matters. Amca had been losing money and decided to sell off some assets, starting with the screw machine division. An old colleague of Phipard's, Art Bancroft, the person who had been in charge of licensing for Continental Screw, bought the patent rights from Amca—with the aid of a loan from Harvey—and set up a licensing business from which the two of them continue to enjoy the royalty flow.

The Lot of the "Company Man"

It's impossible to leave Harvey Phipard's millions without reporting that, in my observation, most often it is the corporation and not the ingenious employee who is enriched by great inventions. Some such "inside" inventors are deeply unhappy with their situations; others accept their lots calmly as part of corporate life.

The discovery of carbonless carbon paper, which was first produced in 1953 by the National Cash Register Company, brought blessed relief from carbon smear to millions of typists. This carbonless system was made possible by a kind of microencapsulation. The trick was to coat oil and dyes with clay and starch and to apply this coating evenly to the back surface of a piece of paper. The dyes were chosen to be on the acid side. The front side of the next sheet was coated with an acid-reactive clay.

When a typewriter key or a pen impacted the front of the paper, the microcapsules corresponding to the shape of that particular writing were fractured on the back of the first page and the ink was physically transferred onto the next page beneath. Bank deposit slips, for example, are made this way.

Credit for the discovery of the microencapsulation system that made this possible goes to Dr. Barrett Green, an R&D man at National Cash Register—but credit was just about all he got. NCR acknowledged the invention one year with a nominal bonus but precious little else.

More recently, Barry Green, upon receiving an award for his breakthrough from the Microencapsulation Society, ruminated, "I appreciate this, but when I think that there are more than a billion pages of carbonless paper produced each year in the United States and two billion worldwide, sometimes I wonder where I went wrong."

Where Barry Green "went wrong"—and of course, he didn't—was not his failure to foresee how immense a business his invention would spawn. His problem was simply that he did not, in advance of the commercial success, seek to negotiate a small royalty from his management. But, unlike Harvey Phipard, he had probably signed his rights away when he entered NCR's employ and most likely felt he was just doing his job.

Art Fry, the driving force behind Post-it Notes at 3M told me that the reward for his role in creating this $100 million business was a dinner praising him in front of his colleagues and "the opportunity to work in this open creative atmosphere." On the other hand, Dan Friel told me that Du Pont had rewarded him "richly" for his inventions and that William Carothers, inventor of nylon for Du Pont, had retired "a rich man."

Two points to tuck away and cherish emerge from the foregoing stories: (1) A key test as to whether or not patents may be licensed as a group, terminating with the "last-to-expire" patent, is whether or not you can reasonably distinguish an end product made under the teachings of one patent teaching from a product made under the claims of another patent in the group; and (2) friendly arrangements with enlightened employers and liberal licensees are the "inside inventor's" best chance. Making a deal with your employer is possible when you pick a good employer. Settling beats suing, nine times out of ten, even though you may have to hire a lawyer first.

From the Employer's Standpoint

You hire good people for your R&D department. It should be no surprise that they have inventive minds. Some of them will make discoveries outside of their assignments. Expect it, encourage it, and provide a basis for compensation at the outset. Good inventions, even simple inventions, are too valuable to waste. Set a policy for what happens with inventions before they happen.

I remember hearing poet Robert Frost speaking at a commencement exercise at Amherst College. He said,

> I tell my students that a fresh idea is the most important event in their lives. I tell them that, if they have an idea during my class, even while I am talking, to reach down under their chair and take a piece of paper and write it down and to give me that paper at the end of the class. I'll

be so interested, I won't even put it in my briefcase—I'll read it on the street as I walk home.

Robert Frost's attitude toward creativity is a far cry from a corporate "suggestion system." Farsighted company executives can structure systems for encouraging invention too. You can read some specifics on this in Chapter 14.

From the Employee's Standpoint

If you are a company employee with an inventive turn of mind, make sure that your job description or contract doesn't include (or, better yet, specifically *excludes*) inventing new products, except where that is your specific job. Keep everything that relates to your own inventions off the company premises. Don't talk about them at work. Don't borrow materials, instruments, books. Unless you have an agreement with your employer that defines what happens between you if you invent something, don't tell your employer about your invention until after you tell your patent lawyer.

What Do You Have to Sell?

When you grant a license for an invention, you give over to someone else the right to make and market products that employ your invention. However, that grant is only the beginning of a buttoned-up license agreement. As the expression goes, the genius is in the details.

I can't make you an instant expert on licensing deals. If you're in a serious negotiation, you still need an experienced lawyer or licensing specialist. But I can alert you to some critical issues to watch out for, worry about, insist upon, or dodge. And, at the end, I'll tell you how to look for and select the people to help you.

What Should You License?

The brief answer to this is "as much as possible." Very simply, if you are making a deal that promises a big royalty stream, you want it to stick. You don't want someone in your licensee's management to look at those checks one day in the future and say to his lawyer, "Find a way out of this agreement." Can such a scoundrel claim that your patent is invalid or ask the Patent Office for a reexamination, even though his company has taken a license? You bet he can.

There are some ways to deter disenchantment. Whenever possible, license more than one patent, even though additional patents cost money. Also, it's wise to keep one or more patents pending on various aspects of your invention because that keeps the door open for still more CIPs. If you can develop a picket fence of patents, it's tougher for the licensee's attorney to tell the boss that all your patents will be held invalid. That's one reason why companies like Lanxide, not to mention most Japanese companies, have such big patent portfolios.

It's also helpful—in my book, almost essential—to license your know-how as well. Now there's a neat line to draw here. The Patent Office requires that you disclose not only how to practice your invention but the best embodiment of the invention. So how can you have any worthwhile know-how left if you've disclosed everything in your patent application?

Well, you can. You may have discoveries that come after the patent is granted. You may have discoveries that do not relate to the patent claims but rather to ways of manufacturing the product, to sources of the ingredients, to making the product larger, smaller, more potent, less toxic, cheaper, tastier, sturdier, and so forth. Now the last thing I would do is to tell you to be less than honest in your patent disclosures—because that could be the basis for losing your patent if challenged. But there is nothing wrong with learning where the line should be drawn between your invention and the know-how needed to commercialize it.

Your patent attorney should be a good source of information on this subject, but you have to be in a position to tell him or her accurately what is already "trade practice" versus what you consider your know-how.

Another valuable property to license is a trademark. If you have actually made and marketed the invention under a trade name, it is invaluable to include that name in the patent license and to state that the licensee shall use it and not market the product without it. The PLAX product sold to Pfizer is a good example.

If you have what you consider a "perfect" trademark but have never sold enough of the product to create buyer acceptance, you should still try to include it in the deal. Experienced licensees may well feel that they can come up with an equally good mark, but if your trademark is registered, appropriate, and apparently clear of conflict, it has a value and helps tie the licensee to you.

Other elements that may deter a licensee from walking away include becoming the source of supply for part or all of the product or sewing up the actual machinery or the molds or the software necessary to produce the product; the business terms of such agreements, however, must be very carefully structured to avoid antitrust and unfair trade practice laws.

Think hard about what you can include under know-how or technology and include in the license agreement a separate "grant clause" for that body of knowledge. Then even if your patent should fail to issue or be invalidated, you still have the basis for a license. In one PRI license, we are able to include under know-how the source of supply for a key ingredient of an antacid tablet—because it was the only place one could buy this ingredient with the required purity and the required low price.

Placing Limits on the License

To maximize your option and royalty income, always consider ways in which you can delimit the rights you are granting. Here are some of the areas in which grants can be circumscribed:

- *By product.* Many inventions enable people to make more than one end product and each product becomes a business opportunity. For example, Alza Corporation's controlled-release drug delivery system, Oros®, was licensed to many different pharmaceutical manufacturers product by product, greatly increasing its gross income and decreasing the risk that one failure would kill the whole system. Sometimes the best way to describe a product limitation is not obvious, but it is always worth careful scrutiny and brainstorming before the license negotiation begins. I know of a case in which a clever toy was licensed to a mass marketer for manufacture in plastic, to a pricier promoter for production in wood, and to a jewelry manufacturer for execution in silver.
- *By territory.* In this era of internationalization of industries, licensees frequently demand worldwide license. There are all kinds of reasons why you should resist this. While a great many companies do business internationally, very few companies are genuinely worldwide in the sense of conducting meaningful marketing in every major country around the globe. Does your prospective licensee have its own presence, a joint venture, or a distributor in each country to be included? What size is its local business now in the product category or a related category? What does its business plan say about its commitment to grow in that country—by how much and when? And, at the bottom line, what minimum royalty or other guarantee of performance will the licensee give you territory by territory?
- *By exclusivity.* An exclusive license? A semiexclusive license? A

sole license (exclusive save for the licensor)? This is easy to think about: The greater the performance required of the licensee, the greater the exclusivity you can grant. If there are high minimum payments, large guaranteed promotion expenditures, and substantial start-up manufacturing commitments, the licensee needs proportionately longer exclusivity.

For smaller companies licensing larger ones, it's useful to study the road taken by such organizations as University Patents, Inc., and Elan Pharmaceuticals PLC. While different in many ways, both companies were founded on the concept of developing and licensing patented technology. Both experienced some success; both went public. But eventually the management of both Elan and University Patents decided that, to realize their real potential, they must make and market products themselves. As Don Panoz, CEO of Elan, told me, "Sooner or later, you have to get more control over your own destiny than a license gives you." That's something to bear in mind when negotiating a license agreement.

Remember that all limitations on the license grant that I have mentioned are interconnected—and must be negotiated concurrently.

Whether you are a corporation or an individual, it's important to get advice on the probable tax consequences of a proposed deal before you go into the negotiation.

For instance, if you license the "entire right, title, and interest" in and to an invention, it's a sale of property and is taxed as a capital gain. Even though it's paid for by royalties on sales over a period of years, it's still an installment sale and you get capital gains treatment. Although, as of this writing, the tax rate on capital gains is the same as the rate on earned income, you may have other capital losses to offset your patent profits. And that capital gains treatment can be available not only to the actual inventor(s) but to those who are "holders," as defined by law, of an interest in the patent.*

If you get too much money from royalties, you may find that it is considered "passive income," and the alternative tax computation will kick in. And no doubt, by the time you read this, the tax law will have been changed again, for better or for worse.

Those are only starters. If you make millions from the mind, and want some of those millions to stick to your pocket, you need an accountant with licensing smarts as well as a lawyer.

*A "holder" of a patent is any person, including a passive investor, who acquires an interest in the invention prior to "reduction to practice."

To Whom Should You License?

There's a short answer to this question: License to someone who already needs what you've invented. Why not just go to the top companies in the field? Because the licensing game is not played on a level field. You know what you've got, but you don't know what your prospective licensee has ready for market or in the development chain. And you may not know a lot of other things, like which person to try to see first, whom to trust, how much to reveal, how much to ask for, and how long to wait for an answer.

There are no universal rules, but experience—my own and a synthesis of other professionals'—provides consensus guidelines and an improvement on guesswork. Here are six clues to choosing the right licensee:

1. *Unless you have a strong reason to go elsewhere, pick the number two, three, or four company in a field.* The top company's executives are likely to feel that they got where they are without you and expect to continue to do just fine. On the other hand, if you go down to number six or seven in the field, you're likely to find that something's missing in the company that your invention can't fix—like enough capital or engineers or salespeople.

2. *Know your prospects' hallmarks.* Companies have hallmarks. Some are known to be open to licensing, some to be uninterested, and some interested only in "going to school." How do you sort them out? Ask around in the trade. Ask other inventors, lawyers, and licensing professionals. And when you talk with a company's top brass, don't be afraid to ask what products they are marketing *now* that came from outside their own R&D.

You can tell a lot about corporate attitudes from the response to your letter asking to present your case. Here, for example, is an excerpt from an actual letter from a big soap company replying to a request for a meeting about a new antiperspirant with a patent pending:

> Because of today's increasingly complex nature of product development and promotion, we primarily rely upon our own organization to create ideas for the development of new products, more useful and attractive packages, as well as more effective ways of advertising and selling our products. In fact, we have many skilled employees working exclusively in those areas.

Catch that condescending tone!

Some companies, such as Gillette, ask people submitting an invention to agree in advance to limit any payment to them to a certain sum, say $100. Score one point for the lawyers, none for new product opportunities.

In fairness to large companies, they are all bugged by hundreds of unsolicited amateur ideas, many of them pretty weird—like that from a woman who proposed a new cookie recipe to a big bakery: The cookie was laced with alum (an emetic inducing vomiting) to teach kids (starting with hers, I guess) a lesson when they raided the cookie jar.

And it isn't always easy to separate the wheat from the chaff; it's easier to say that you didn't read the suggestion and send it back.

Does that mean you can't do business with these people? Not necessarily. It simply means that you may have to establish contact through intermediaries—suppliers, engineering firms, bankers, advertising agencies, anyone doing business with your target company. I once got an appointment to see the CEO of a big company by sending a sample of my product through the mail to his wife at his home address.

3. *Having secured an appointment, be prepared to sign whatever secrecy agreement they want.* But also be prepared to reveal only as much as you absolutely must until it's established that they are interested. Then, when they ask for details, ask them to sign *your* secrecy agreement. At the end of this chapter, you'll find copies of confidentiality agreements that I have found acceptable to almost all companies. One is to bind the company or the person receiving the confidential disclosure; the other is for mutual confidentiality for use where information may be disclosed in both directions.

4. *You should almost always go to more than one company with your presentation.* The right number is probably three meetings held as close together as possible. Ideally, you will tell each company that you are presenting to only two or three marketers essentially simultaneously and offering identical terms. As my friend Gene Whelan put it, "The first girl who says 'yes' gets to dance." If you don't have multiple concurrent targets, what will you do if you get a letter with words like these: "Even though the Marketing people were interested, they seem to have too many other concepts they were looking at and just couldn't seem to get at it. The R&D group for this particular area has recently changed and Marketing was hoping things would happen on this but since we can't give you any time

frame in which this might happen, we are returning the samples." (This is an actual quote—would I make up something like that?)

5. *Get all the intelligence you can on what your target companies are doing for themselves.* Deduce their strategic objectives by scanning back copies of *The Wall Street Journal, Business Week, Forbes,* trade journals, and other sources. Find a friendly face at a company that you don't intend to license and have a chatty lunch. Check out possible objections to your presentation and get the scuttlebutt on what's happening with competitors. Interview an ex-employee of your target company: How did management view its recent successes and failures? Who were the villains and heroes? All such homework pays off in dollars if you get to a serious negotiation.

6. *Loosen up your thinking regarding your "best licensee."* Robert Goldscheider, Chairman of the International Licensing Network, was seeking a licensee for a patented plastic chain. Big makers of stainless steel chain, headed by Rexnord, dominated the market, especially for conveyor belts. The new plastic chain was clearly a threat; it was lighter, quieter, cheaper, and needed no lubrication. But Bob wisely ignored the steel chain makers with their massive plant investments and deep commitment to the merits of steel. Instead, he identified a smaller company that produced plastic rollers for conveyors and already called on the best potential customers: food plants, dairies, and bottlers. Goldscheider was right; he made a profitable deal for all concerned within a matter of weeks.

The Point of Entry

In his book *The Terrible Truth About Lawyers*, Mark McCormack wrote: "It is an absolutely basic rule of conducting business effectively that to get a favorable result, you've got to get to the decision maker." Who's going to argue with that?

The problem, of course, is to figure out which person, holding which title, has the imagination and drive and authority to license a new product and hang in and become its champion.

You start with quite a choice. Here are some of the titles of people who have power over licensing-in inventions: chairman ... president ... chief executive officer ... vice-president marketing ... vice-president research ... vice-president manufacturing ... vice-president corporate development ... vice-president licensing ... vice-president engineering ... new product planning manager ... manager, new product development ... product manager, new products ... director, planning and development ... director of licensing ... advanced engineering and development director ... manager, process development ... market development manager ... product manager.

Of course, what counts is not just the title but the individual. So getting to know what kind of person carries any given title is all-important. Since that's not always possible, try to remember some truisms about corporate structure, like "enthusiasm filters down, not up"; or, "if you have to go up the approval ladder far enough, sooner or later you'll hit a missing rung."

Short of real knowledge as to the best point of entry for starting a licensing discussion, go to the top. Write to the president or the chairman of the company.

As you know, in most companies the president or the chairman doesn't see the letter until after his or her secretary has read it. I've used two methods to ensure that this person puts my letter in the stack for the boss to read and not the circular file.

One is to attach a small handwritten note addressed to the secretary personally explaining why my invention or product will interest the top executive. The other is to begin the letter by talking about the prospective company rather than by describing my innovation—for example, "Last year, XYZ Company had $335 million in sales, but, as you know, even though that was an increase, profit margins were hurt by price cutting in the market. I want to offer you a product that has wider margins than your present average—a product with strong patent protection that can be ready to sell next year." Whatever. Just show that you have studied your prospect's company and needs. It won't always help, but it's way ahead of starting out by explaining how great you think your invention is.

Other times, I've had the president of a company buck my letter down to someone with a note on it, "Look into this." That downward flow may ruffle some feathers, but even if it does, that person you're referred to still has to report his or her findings back to the president, and you can still follow up with the prez as circumstances require.

Another way in is to call the president's secretary and ask for her help in meeting the president or finding the right person to approach.

Alternatively or concurrently, check every business directory in the library for the names of all the prospect companies' top officers and their board members; see if any name suggests a point of contact. Go to trade shows and conventions, where you can often meet people more easily than you can corral them in their offices. Get the current news about what's going on inside your prospect companies, their winners, their losers, their good guys, and villains.

Finally, what kind of letter should you write to open the door? It's my belief that an initial letter to a top executive should be reasonably short. Here's what you should remember:

• Put the headline in the first paragraph. If you have more to say, attach exhibits. Letters longer than four paragraphs seldom get read by CEOs.

• State succinctly the *principal* advantages of the invention (1) to the user and (2) to the company you're writing to.

• State at once that your development has an issued or pending patent plus other know-how.

• State your credentials as a company or as an inventor or as an entrepreneur—or all three.

• State that you want an initial nonconfidential meeting at which you can present more detailed information.

Having made so many points about ways to push into the corporate organization, I must, in all fairness, report that some gifted inventors have the knack of getting the potential licensee to come to them.

You do this primarily by making yourself an expert on something and putting yourself in front of the world through industry meetings, speeches, reports, and personal contacts.

An absolute classic in this category is Maurice Hiles, a Britisher now living in Ohio. When Hiles was working in England on an Anglo-American medical research program, examining the problem of shock waves transmitted to the human body when the heel strikes the ground, it was found that scientists had been measuring the shock in the wrong place, that you have to measure the shocks in the bones and not the tissue. Because Hiles was the materials scientist on the team, he was given the task of finding a shock-dissipating material. And his first step was to bore some little holes in his own legs and put in some computer sensors so that shock to the bone could be measured. As Maurice Hiles told me,

> The tests showed such a striking change between the soft tissue and the bone itself, it hit me that the calcaneal fat pad in the heel must have some extraordinary properties to let us walk at all. So we analyzed it and found a series of compartments with little fat globules in them. I thought, if I can simulate this on a microscale in plastic, we could have a very efficient energy-dissipating system. That material is

> called Sorbothane®. In fact, it's now manufactured in the United Kingdom, the United States, Brazil, Japan, South Africa, and Australia.

Suddenly, I realized that it was Maurice Hiles who had made possible the tennis shoes that let me continue to bounce around those hard-surface courts despite my worn-out knee joints. I asked Maurice how he had found a licensee for a discovery that might have ended as nothing but a scientific curiosity. He replied:

> "I didn't. They found me. I was a consultant to the Allied Polymer Group and they were taken over by BTR, one of the largest fabricating companies in Europe. I actually made the deal with help from the British government, the National Research Development Corporation, which is now called the British Technology Group. All along in my history, serendipity has sat on my crown."

Not everyone can depend upon serendipitous good fortune, but following Maurice Hiles's path of aiming inventions at known problems and then associating oneself with reputable technology groups is surely a pathway to explore.

Whether a licensee comes to you or you reach out to find the right company, eventually you'll come to the point of presentation and negotiation—where the battle begins in earnest.

Permutations of Presentations

Before that first meeting, find out who is going to be there to represent the prospective licensee. If it's an informal meeting with the company president, go alone or with one other person. But if, as is the case with most larger companies, they bring out the troops—marketing people, R&D people, licensing people, and so on—it's important *not* to be alone.

Ideally, you need a team with expertise corresponding to the talents they are assembling: a business negotiator, a lawyer, a technical expert, a marketing maven, a production specialist. This may not be easy or always appropriate. In any case, you need one or more additional people on your side whose function is to watch and listen to how people are reacting to the presentation, to share the presentation burden, and to answer questions.

Regardless of the number attending the presentation, you have to decide if the real decision maker is there. If that person is missing, you have another major decision to make: Should you leave copies of your presentation and/or samples behind to be bucked upstairs, or should you say politely but firmly, "Thank you very much for coming. If you folks like this, we'll be glad to come back and present it again to the boss—not that you aren't perfectly capable, but we have found that when questions arise, it's best to answer them on the spot."

Arguments can be made for a variety of presentation techniques. The best ones I've seen included either an actual demonstration of the invention in use, preferably something that the people at the table could do themselves, or a videotape of the product in use or even a possible commercial for the product. However you do it, dramatize the invention because if you don't generate enthusiasm, natural corporate inertia will grind matters to a halt.

In addition to dramatizing the product/invention, you must tag at least four bases for a successful presentation:

1. *Provide evidence that the invented product fills a genuine need.* Show that there are people out there who will cheer its introduction and buy it. This does not have to be a national study by a renowned market research firm; most large companies have their own prejudices as to what kind of major research they accept. But if you can report that five store buyers said they liked it or that ten company purchasing agents agreed to put in a test unit or that thirty consumers tried it at home and actually paid for it—well, at least it puts the doubters on the defensive.

2. *Analyze the probable tough questions in advance of the meeting.* Where are the weaknesses or unknowns? What is the cost of goods? The strength of the patent position? The proof of technical superiority? The size of plant investment? How about durability, stability, material availability, regulatory or environmental issues, packaging, and so forth?

It's almost always better to bring up the possible "killer questions" yourself and answer them at the meeting—because if you don't someone will raise the question after you've left, when there's no one to defend your side.

3. *Rehearse the presentation with a devil's advocate.* Bob Goldscheider tells the story of a man who was negotiating a license with a South African steel mill owner who was known for his rough tactics. Bob's client was a pretty tough guy, too, built like a pro football player, with a negotiating stance to match.

At the dress rehearsal, Bob played the mill owner. As the discussion went along, suddenly he threw the whole contract at his client, saying, "This is a terrible contract. Why are you wasting my time?" Bob's client almost exploded.

"Wait a minute, hold on," said Bob, "This is only a rehearsal. Now what will you do if this should actually happen at the negotiation? I'll tell you what to do. Nothing. Just sit there and put the contract back on the table and wait for him to make the next move."

Goldscheider had prescience. It happened almost exactly as rehearsed, contract thrown across the table and expletives pouring forth as to how bad it was. Bob's client put the contract back on the table, sat there for a moment, and then said, "Now, shall we talk?" The tension broke, they worked out the deal and went on to a very pleasant lunch. Rehearsal pays.

4. *Control your exposure.* Even though you have a signed confidential disclosure, feel no compulsion to tell your prospective licensee everything you know until the ink dries on the contract and the check clears the bank. What you disclose is a function of the other side's absolute "need to know," and that's a matter of judgment, yours as well as theirs. The major principle is to reveal, indeed prove, what the invention does, but not necessarily every detail about how it does it or how you manufacture it or where the materials come from or a dozen other items.

No matter how pleasantly the presentation and negotiation may be going, never forget what Yogi Berra said: "It ain't over till it's over!" That means, protect your secrets and protect your product at every stage, consistent with what must be revealed to make the deal. Do the Dance of the Seven Veils. And that means throughout the presentation, the negotiation, the option period, and sometimes even after the license is granted.

A classic case in point is an option/license I wrote with an international pharmaceutical company in which we provided samples from our own laboratory that the company agreed to test (*in vitro*) but not analyze. If the samples met their standards, then they would exercise the option and take the license and we would give them the detailed formulations and manufacturing procedures. We limited our exposure; they limited their financial risk. The test results satisfied them that the product did what we had claimed it could do; and they paid us a substantial sum to take the license, a sum they would never have paid without doing their own firsthand evaluation.

Making the Best of the Unexpected

Sometimes you can do everything right, or feel that you have, and the deal, like Humpty-Dumpty, still comes tumbling down. When it does, it pays to look hard at the technology on which you've lavished all that time and money to see what's needed to give it a second life.

I had that happen to me with a wonderful aerosol invention. Imagine a normal-looking aerosol can; you squirt it into a glass of water and you immediately have sparkling club soda. Put flavor in the can and you have orange soda or cola or whatever.

This was the discovery of a delightful man named Norman Ishler. He uncovered it serendipitously (again!) when an experiment he was doing for General Foods went wrong. He discharged an aerosol can designed to expel a simple food additive into a beaker of plain water; within moments, tiny bubbles began rising through the liquid. And they just wouldn't stop.

Instead of trashing the experiment as a failure, Ishler asked himself why it had happened and found out how to replicate it. The reason for the bubbling turned out to be quite simple: The propellant used was a fluorocarbon, which is heavier than water; the fluorocarbon sank to the bottom of the glass and formed bubbles, which then, one by one, propelled themselves to the surface.

As PRI looked into the uses for this technology, we concluded that, even though an argument could be made for economy, it was too hard to buck the big soft drink names and to educate people to this new way of making their own.

A "bubble bath" sounded like a natural application. We made samples, dubbed the product Great Feeling, and made an immediate sale to Sterling Drug Co. Even while the license agreement was in preparation, samples of the product were furnished to Sterling's technical people. But the day before we were to sign the license and pick up the check, we received a call from Bill Heicke, Sterling's vice-president for licensing, telling us that their scientists were afraid that someone taking a bath in our Great Feeling might fall asleep in the tub, slip down to where her nose was near the water, and inhale our fluorocarbon bubbles . . . and die. How could they go ahead with a product that might kill people? No way!

Recovering from this blow, the PRI crew thought hard about what else would make a perfect bubble-aerosol product. I said, "Whatever it is, make sure that it's as far as possible from the nose." To which my partner, Gene Whelan, immediately replied, "A foot bath!"

We licensed that product to Pharmacraft Corp., and Ike McGraw, the president, loved it and the product manager did a great job producing a commercial with tired, end-of-the-day feet being rejuvenated in a transparent bowl showing the bubbles tickling the soles and toes. Consumer tests were gangbusters.

Then, a few months before national launch, the Environmental Protection Agency banned fluorocarbons on the grounds that they were destroying the ozone layer above the earth. And the alternative propellant, hydrocarbons, being lighter than water, wouldn't sink to the bottom and create those nice bubbles.

I'm still thinking about that one.

* * * * * *

Finding the right licensee and making the sale is half the "easy" road to money. Nailing down the financial details and making sure that the licensee's lawyers don't take away from you in contract what you've won in the presentation is the other, indispensable half. We'll follow that yellow brick road in the next chapter.

Appendix A. Letter of Agreement for Invention Owner Making Disclosure

[*Date*]
[*Name of Individual*]
[*Name of Company*]
[*Address*]

Dear ________:

We are in possession of information relating to our [*name of product or technology or invention*] (hereinafter referred to as "the Subject Matter"), which we consider confidential. We understand that you are willing to accept such information from us for purposes of evaluation and determination of your possible interest therein. You agree to receive such information from us during the term of this Agreement on the following basis:

1. Under this Agreement, you will hold in confidence any and all technical information on the Subject Matter disclosed to you by us, in writing, and marked "Confidential," except:
   a. information which at the time of disclosure is in the public domain;
   b. information which, after disclosure, becomes part of the public domain by publication or otherwise, except by breach of this Agreement by you;
   c. information which you can establish by competent proof was in your possession at the time of disclosure by us and was not acquired, directly or indirectly, from us; and,
   d. information which you receive from a third party, provided, however, that such information was not obtained by said third party, directly or indirectly, from us.
2. You agree that you will not use the technical information relating to the Subject Matter, which you are required hereunder to keep confidential, for any purpose other than the aforesaid evaluation and determination of interest, without first entering into an agreement with us covering the use thereof.
3. Your obligation under this Agreement shall expire on the fifth anniversary of the date thereof. We agree that, after said fifth*

*Five years is typical, but this is a negotiable point. An agreement without a time restriction is called "evergreen" and is seldom acceptable to the restricted party.

anniversary, we will make no claim against you or any of your subsidiaries with respect to this Agreement, except for patent infringement.

If you agree to the foregoing, kindly indicate your acceptance thereof and assent thereto by signing and dating the duplicate copy of this letter in the space provide for below, returning such signed copy to us. We will then proceed to disclose to you our information relating to the Subject Matter.

Very truly yours,

[*NAME OF COMPANY OR INDIVIDUAL*]

By: ____________________

ACCEPTED AND AGREED TO:

Company: ____________________
By: ____________________
Title: ____________________
Date: ____________________

Appendix B. Mutual Secrecy Agreement

The parties hereto are:

[Your company or individual name]
[Address]
[Their company or individual name]
[Address]

Both parties have agreed to make available to each other certain confidential, technical, and marketing information and product samples relating to the [*name of your invention or technology* and *name of their product or technology*], so that the parties may discuss and/or pursue the mutual development of new or improved products.

Both parties agree that with respect to any confidential information disclosed by one party to the other, providing that said information is disclosed in writing marked "Confidential" (or confirmed in writing marked "Confidential" within thirty (30) days of oral disclosure), the receiving party:

(a) Shall use said confidential information only for the purpose of determining its interest and/or capabilities as aforesaid.
(b) Shall disclose the confidential information only to those employees or consultants (who are personally committed to maintaining the information in confidence) necessary for it to make said evaluation, and shall in no instance disclose the same to any other party for any purpose without the written consent of the party providing the information.
(c) Recognizes the trade secret nature of the information.
(d) Shall have no obligation of confidentiality with respect to information that:
    i. is in the public domain by use and/or publication at the time of its receipt or enters the public domain thereafter through no fault of the receiving party; or
    ii. was already in its possession prior to receipt as shown by written documentation to be delivered to the disclosing party within thirty (30) days; or
    iii. was properly obtained from a third party not under a confidentiality obligation to the disclosing party; or
    iv. was previously developed, independently, by the receiving party, as shown by written documentation to be delivered to the disclosing party within thirty (30) days.

Unless the parties shall otherwise agree in writing, the parties shall return all confidential information provided by one to the other within one (1) year from the date the confidential information was received.

This Secrecy Agreement does not constitute a commitment nor an obligation on the part of either party to enter into any other binding contractual Agreement.

The obligations of confidentiality under this Agreement shall be limited to a period of five (5) years from delivery of the information.

This Agreement contains the entire understanding of the parties hereto with respect to the matters herein contained, and supersedes all previous agreements and undertakings with respect thereto.

IN WITNESS WHEREOF, the parties have caused this Agreement to be executed by duly authorized officers.

[*NAME OF YOUR COMPANY OR INDIVIDUAL*]	[*NAME OF OTHER COMPANY OR INDIVIDUAL*]
BY: ____________________	*BY:* ____________________
TITLE: ____________________	*TITLE:* ____________________
DATE: ____________________	*DATE:* ____________________

Chapter 11

Licensing Agreements: Getting Yours and Keeping It

When you're about to entrust the future of your baby to foster parents via a license agreement, you have to look very carefully at both the written document and the people who will be charged with implementing it.

It's commonly accepted wisdom that the licensor's best assurance of sustained success is to have someone inside the licensed company as a champion. Wisdom it is, but automatic it is not. You have to look for this person from the moment of your first meeting. Consider who is in a position where his or her career could benefit immensely from the success of your invention. Support that person with useful information, praise that person to his or her superiors, get that person tickets to Carnegie Hall or the Super Bowl, whatever it takes.

At the same time, try to identify your possible opponent within the company and neutralize that person, especially if it's an R&D person. How do you do this? Be inventive; there's no single answer.

At a meeting at Bristol-Myers, my team was presenting a new antiperspirant technology, a powder that briefly turned into liquid when you applied it and then dried again in moments. I kept watching their second-in-command R&D guy, Len Mackles. As the presentation continued, he twisted in his chair, he doodled on a pad, then almost visibly withdrew behind his thick eyeglasses. Now I knew this was a very bright man; I couldn't ask him what was bugging him, but I could guess. He was an expert in the antiperspirant field and it was likely that our invention was close to something he'd tried before.

So when we negotiated the option/license agreement, I put in a clause that, if the licensee had prior knowledge of our invention, their people must, within thirty days after signing and disclosure of our technology, provide us with written evidence. In that event, the agreement would terminate and they would get their option payment back. The thirty days crept by and we heard nothing. Not only did we breathe easier, we set out to make Len Mackles our ally and to add whatever he knew to the product pot. Not until two years later, when Len Mackles retired from Bristol-Myers and became my valued colleague on other projects, did I find out that he had lab notebooks showing almost the same discovery but stopping short of reduction to practice. Without that thirty-day clause, the deal would surely have come a cropper.

How Much, How Soon, How Sure?

Setting the value of a license for an invention is one of the more difficult price negotiations in business—because you're dealing with a parlay of variables, all the way from what it will take to produce the product at a given cost to what will be needed to sell a projected volume at a certain price.

So the first task is to squeeze out the doubt where you can. If the product is already being made and sold in one country or one field, some answers exist or can be postulated; if not, you have to move through option periods while the unknowns are being resolved by yourself or your licensee. Thus Pfizer, Inc., paid a Japanese pharmaceutical company $30 million to acquire an exclusive U.S. license to a drug that was already a great success in Japan.

In contrast, Pfizer paid Tom Thornton only $25,000 for a six-month option on his not-yet-marketed Superfloss®, extended it for another six months at the same price, and proposed another extension on the same terms. Tom walked away and ultimately made an interesting deal with Cooper Laboratories, details of which you'll find near the end of this chapter.

Moral: More precious than gold is time. Don't let them test your invention to death or extend options forever.

Unfortunately, many really talented inventors are not good businessmen, especially when it comes to getting paid well for their discoveries. Worse yet, many good inventors don't recognize this or don't admit it in time to make deals that would make millions for them.

Jack Rabinow, a prince of a man with a beautiful mind, told me that he never really became rich; in fact, monetarily, he just squeaks into the qualifiers for this book. Now this is the Jack Rabinow who invented the straight-line phonograph tone arm, the magnetic-particle clutch for automobiles, the letter-sorting machines used by the U.S. Post Office, the self-regulating clock, and more. Jack is also the man who showed me his recent invention, the first door lock that no burglar can pick (honest!).

There are two reasons why Jack Rabinow is more famous for his ingenuity than for his wealth: (1) like many inventors, he is so focused on the process of inventing that making money gets a small share of his mind; and (2) again like many other inventors, he never found a financially skilled business partner who could make the deals that make the big money.

How High Is Up?

How do you arrive at the right price for a license? How do you pinpoint the terms that will induce a businessperson to risk his or her capital, talent, and reputation to commercialize your innovation, discovery, or technology?

At the most rudimentary level, it's what the traffic will bear. Some years ago, a man named Bernie Kahn was starting out in the new product ideation business. He and his partner had a date with a prospective client who had indicated a serious interest in retaining them. Bernie's partner said, "How much shall we ask for?" Without hesitation, Bernie replied, "I'll ask for $2,000. Then, watch him. If he doesn't flinch, you say, 'per month.' And if he still doesn't bat an eye, I'll say, 'each.' "

At a more sophisticated level, each situation has many complicating factors that require both short-term business savvy and long-term financial foresight. While it is easy to say, "The inventor is entitled to one-quarter to one-third of the profits that the licensee will make from the invention," the fact is that corporate management does not want to reduce average profit margins, even if total profits increase. As a result, royalties are treated as a cost and are often multiplied by the normal markups of the manufacturer and the distributors, resulting in a premium price for the product. It can be argued, and you should when you can, that a unique product will need less advertising and promotion or discounts, that a patented product will hold off competition longer, and that a licensed-in product has incurred less corporate overhead than internally developed products. So the real net profit margin is usually not reduced by the royalty. The financial officers of large corporations may not accept this concept, but net-profit-minded CEOs sometimes will.

The bottom line of all royalty negotiations is to find a point that (1) lets the licensee make money as soon as possible, (2) provides an incentive for a major initial marketing effort, and (3) delivers total royalties to the licensor that, in effect, make him or her a "partner" in the licensee's business.

Statesmanship in Licensing

The story of how recombinant DNA was commercialized is both fascinating and illustrative of how a complex licensing problem can be worked out. It is also a good example of a university achieving major funding from its discoveries rather than just from its alumni.

While enjoying a corned beef sandwich at a delicatessen at Waikiki Beach in Hawaii in 1972—during a break from an international convention on microbiology—Stanley Cohen of Stanford University and Herbert Boyer of the University of California not only broke bread together but broke through a barrier in bioengineering. Boyer had worked on advanced technology in the use of "restriction" enzymes to "cleave" DNA at a preselected site. Cohen had been experimenting with plasmids, covalently closed molecules that could be used to carry new genetic matter into a living organism. The science of bioengineering was born when, within four months, Cohen and Boyer produced the first recombinant DNA.

Word of the discovery reached Niels J. Reimers, director of Stanford University's Office of Technology Licensing, in May 1974. By that time, Boyer and Cohen, together with other scientists, had already asked the National Institutes of Health to issue safety guidelines to allay fears that another Frankenstein's monster or a dread disease would be created in some laboratory; and they had written an article for the *Proceedings of the National Academy of Sciences* (*PNAS*), which was scheduled for publication in the fall.

Reimers promptly approached Cohen and Boyer about filing for a patent. Cohen was diffident, saying that it would be twenty years before anything commercial would be accomplished. But Cohen also said he would go along if Boyer wanted to do so. Reimers moved forward. He called Josephine Olpaka, a licensing officer at the University of California, and quickly made an agreement that the two universities would share the income fifty-fifty but that Stanford would receive 10 percent off the top for selling and administering the licenses and 5 percent more for expenses.

From that starting point to the introduction of the first genetically engineered product—a human insulin engineered by Genentech and marketed by Eli Lily—Reimers faced a series of complications:

• The American Cancer Society, the National Sciences Foundation, and the National Institutes of Health had all made grants to Cohen and/or Boyer. Thus, each might claim a stake. Friendly persuasion brought agreement that the invention could be administered "on behalf of the public" under Stanford's existing institutional patent agreement with the NIH.

Point for inventors: Whenever you take money from anyone, whether it's an individual or an organization, if you ever expect to get personal rewards from it, read your contract carefully.

• Between the time needed for sponsor sign-offs and the time consumed in proper preparation of the patent application, Reimers's group was barely able to get the filing to the Patent Office a week before the deadline when, by reason of the publication in *PNAS*, it would have been barred under the one-year statutory limitation.

Point for inventors: Scientists who feel they have a duty to publish, or inventors who feel compelled to make speeches, should check with their patent counsels before the public event.

• While the Boyer/Cohen patent was pending, an application by Ananda Chakrabarty, a scientist at General Electric Company, for a new bacterium that would chew up oil spills brought the issue of the patentability of "invented" living organisms to the attention of the scientific community. There ensued long and sometimes vitriolic debates as to whether or not recombinant DNA research would be opening a Pandora's box. After heavy discussion, scientists reached a consensus that it would be proper to continue DNA research under NIH safety guidelines. Stanford University officials worked within their own research community and then with government officials, including congressional subcommittees, to reach an agreement on ground rules for how genetic engineering patents might be administered with regard to the commonweal.

Point for inventors: During a patent's pendency is no time to sit still but rather an important opportunity to gain allies—licensees, of course, but also anyone else who can help you, from government officials to production engineers to members of the fourth estate.

• Realizing that they had a patent application which, if granted, would give them power over the entire field of genetic engineering, the Stanford people decided early on to license nonexclusively. Opening the door to everyone, however, was almost a sure invitation to a lawsuit from someone, or at least a demand for reexamination. To help avoid the latter, they decided to waive the secrecy of the patent file and to open it to public examination so that third parties could make comments even while it was being considered by the patent examiner. This full and free opportunity for anyone interested to review the facts, they felt, would later provide a strong presumption of the validity of the claims.

Point for inventors: Stanford's statesmanlike move was also an economically sound one. Considering the long time line for development and approvals of genetically engineered products, it would have been disastrous to wait for the patent to issue and then be faced with legal actions. The first Cohen/Boyer patent did, in fact, issue December 2, 1980, and, because Stanford filed a "terminal disclosure"—a Patent Office procedure under which the applicant agrees that all continuations or divisions of the original patent filing shall have the same expiration date—all coverage will expire in 1997. This is just one example, and a good one, of how business planning must be an integral part of the patent process.

- It was in setting the licensing terms, however, that Reimers and his team shone brightest. The major considerations in determining royalty policy were, first, that because application of the discovery was necessary before any major products and royalties could be generated, they wanted as many licensees as possible to be working at it; and second, that because there were many different kinds of applications, different royalty arrangements would be needed for base genetic products (used to genetically alter or manufacture something else), process improvement applications (enhancing productivity), bulk products (something used as part of, or integrated with, another product), and end products (something to be sold "as is" to the end user).

The royalty plan flexibly met all these constraints. It called for a mere $10,000-per-year minimum advance payment by each licensee. Even the smallest company could afford that. Royalty rates were scaled from 10 percent on basic genetic products down to 0.5 percent on large volumes of end products. To hasten sign-ups, any company signing a license within the first four months was offered five times the $10,000 advance fee as a credit against future royalties. Almost every imaginable detail was covered in the original printed license offering, and seventy-three companies signed within the initial four months.

Point for inventors: Every invention licensing situation is unique, and success or failure often rests upon creating terms with incentives to sign up promptly, with later payments scaled to various foreseeable events.

How successful was the recombinant DNA licensing program? Dr. Floyd Grolle, who administered the licensing program, told me that receipts from the fall of 1980 through August 31, 1990, totaled $21,891,000. Fiscal 1991 royalties hit $17 million by virtue of one special payment, and every year until 1997, when the patents expire, is expected to average $11 million. The total: close to $100 million.

Stanford's standard arrangement calls for the university to receive one-third of net patent licensing income, the school department to receive one-third, and the inventor the remaining third. So even though the recombinant DNA discovery was divided between Stanford and the University of California, everyone concerned was well rewarded. Niels Reimers set a high standard for universities to follow in organizing and running a technology licensing department.

And Stanley Cohen and Herbert Boyer can afford to go back to Hawaii for another corned beef sandwich—or even to buy the delicatessen.

A Licensing Map and Compass

Although every licensing situation is individual and requires imagination in negotiation, there are some practical guidelines that may help you to make the deal happen and to protect yourself at the same time.

Setting the Financial Terms

Some industries have norms. For years, a 5 percent royalty was almost standard in the toy industry. For pharmaceuticals, royalty rates range from 3 percent to 15 percent. Although license agreements are not normally public information, check with licensing officers or lawyers in the relevant industry to find out what's in the ballpark.

As mentioned earlier, one widely quoted licensing ground rule is that the invention owner should be paid between one-quarter and one-third of the incremental income that flows to the licensee from the use of the patent or know-how. What is meant by "incremental income"? It can be savings made possible by applying the invention to the production of an existing product or it can be added profits from the sale of a new product.

The former is fairly easy to calculate. But "profits," of course, are a function of what the licensee decides to throw into the cost of goods.

Also remember, while industry "royalty norms" are always useful guides, norms usually mean averages—and wouldn't you want to be on the high end rather than the low end of the averages? So in setting the royalty, look closely at your product's profit margin versus what you figure out to be the prospective licensee's typical margins; if you're raising the profit levels, you should be entitled to a bit of that increase. Don't hesitate to ask you prospective licensee to share with you his "target" percentages for manufacturing costs, equipment amortization, marketing costs, general overhead, and so on. If you compare your projections for your new product with these target areas, you can make a more intelligent licensing proposal. If your prospect won't share such information, use industry averages and be aware that you're dealing with people at a long arm's length.

Similarly, you can strengthen your negotiating position by looking at the incremental effect on the licensee's entire business. Check how your project compares with the licensee's overall return on investment. What will it do for the licensee's share of market? And what would be the potential effect on the prospective licensee if your invention were licensed to a competitor?

By doing a thorough analysis of such factors, you create tangible arguments. You can use them not only in negotiating the basic royalty but in offsetting unpalatable points that may be advanced by the licensee's negotiator.

"At the end of the day" (as our British friends often say when introducing a killer point), as much as rationality counts in setting the financial terms for an invention license, negotiation is an art. The emotional temperature of your potential licensee is a function not only of greed for new profits—nominally the stuff that deals are made of—but of a whole gamut of little fears: fear that the new product will fail, fear that you'll really sell it to a deadly competitor, fear that some other guy in the company will seek to sink the product *sub rosa*, fear of becoming the product's champion and getting fired if it fails. Reading that emotional temperature and playing on it is the mark of the truly skilled negotiator. If you, the invention owner, aren't good at it, find someone who is and make him captain of the team.

Front Money

Where the two sides most often have trouble reaching a happy agreement is on the payment of money up front at the time when the agreement is first signed. Naturally. The licensee would like to pay everything out of the profits on sales. The licensor would like to have some reimbursement for the time and money already invested in the development, especially for an exclusive license.

From the invention owner's point of view, there are additional compelling reasons why a large sum of money at the outset is vital, particularly for an exclusive license. From the date the agreement is signed until the first sales of the product, many things can happen. People inside the company may be working on alternatives to the technology. Others may be pushing projects that compete with yours for plant investment or marketing dollars. Management may change, strategic objectives may change, the company itself may be bought and the new owner walk away from you.

In particular, licensors of inventions that generate relatively small sums of money always run the risk that it may not be worth the attorney's fees to enforce the license agreement. That's what happened to Dr. William Waters. He invented an air-cooled pith helmet, licensed it to a small company, and saw it make a modest market success. Then a larger company gained control of his licensee and the new owners promptly told him that his patent wasn't applicable to the helmet they were making. You can write the best license agreement in the world and you still can't control people's behavior.

These and other perils are reason enough to fight for enough cash up front so that (1) executives will think twice about kissing it off, and (2) you will have enough cash to keep improving what you've licensed and enough time to work as closely as possible with the licensee to keep your invention moving to fruition.

The same underlying principle obtains with respect to guaranteed minimum royalties or other techniques for ensuring that the licensee will put real effort into marketing the invention, such as guaranteed advertising expenditures or a specified number of sales calls. Absent a contractual spur, products licensed from "outside" can find themselves low man on the budget totem pole.

Beating the Lawyers

Many company lawyers see themselves like the hero in my high school Latin book, a staunch figure at the bridge, sword in hand, protecting the turf on the other side against all comers. The legend under the picture, as I recall it, read "Horatio pontum defendit!" You may have met lawyers like Horatio.

Well, you can't blame the lawyers for everything. You can go back and find plenty of cases where the lack of proper legal counsel cost companies money. No cautionary tale is more pertinent than Warner-Lambert's licensing of the formula for Listerine®, a bit of history recounted to me by Warner-Lambert VP Joe Carpino.

The Listerine "license agreement" was handwritten in 1881 on a single piece of paper and signed by Dr. J. J. Lawrence, who invented the formula, and by Jordan Lambert. The agreement called for the payment of $20 per gross of the original containers, which, I believe, ultimately translated to about five cents per bottle.

Unfortunately for Warner-Lambert, the license did not specify a termination date. So the company has now paid royalties on Listerine for over a hundred years. Fortunately for Warner-Lambert, however, and unfortunately for the licensor, the agreement called for a fixed royalty, not a percentage of sales, so that the royalty cost per unit has continually diminished through the years.

These days, company lawyers strive mightily to protect their bosses against every contingency. That's what they're paid for and it's certainly their only chance to chalk up brownie points. In a deal of any substance, you, the invention owner, need a lawyer who is just slightly smarter than corporate counsel.

There is a really red flag to watch for in negotiations. That's when the licensee's legal eagles start to change the terms to which you and the businesspeople have already agreed—or to kill the deal altogether. I recall, for example, one lawyer at Pennwalt who insisted that the company couldn't sign an exclusive license because of potential antitrust problems. When I pressed him as to exactly what problems concerned him, he dodged away. It was finally agreed that we would make it a semiexclusive license, but he then wanted to reduce the payments.

When you come up against someone playing that game, you have to look at your hand and decide if you want to fold or if you want to bet the whole thing and get the other guy's cards on the table. It helps if you have an ace in the hole—such as a strong relationship with the company president or another licensee in the wings. In that case, we licensed somebody else.

All in all, the serious problems in working out agreements with company lawyers arise when they try to change the business terms. Nor are such changes always minor. Twice I have had the licensee's attorneys add clauses to a draft license limiting the total amount of royalties that could be paid. Granted that this could be a negotiating point, albeit a tough one, it belongs in the business discussions, not in the license language fight.

There are two crucial moves to make in order to avoid losing the battle to the lawyers after you've won the deal with the businesspeople.

First, when the presentation is completed and your prospect says, "What's the deal?" hand them a written outline. I use a form that the British call Heads of Agreement; it's a way to get the essence of the agreed business points on paper while these points are fresh in everyone's mind and without waiting for the lawyers to cover reams of paper. An outline of typical items for Heads of Agreement—specifically for a licensing contract—is provided at the end of this chapter.

Second, have *your* lawyer write the first draft of the actual agreement. That's much easier than fighting to eliminate unacceptable clauses or language in which the company's lawyer has an emotional investment. Ralph Levine, vice-president and general counsel at Carter-Wallace, Inc., and a charming man personally, gave me all kinds of trouble on this subject; he insisted on using only his form of agreement—which was almost totally different from any other license agreement that I or several of my attorneys had ever seen—on the grounds that Harry Hoyt, the president and principal stockholder of the firm, knew how to skim through this form and wouldn't sign anything else because he'd have to read the whole thing. That was probably true, although I know that Ralph used the ploy to get his pet points embedded and we fought hard to get a few of them changed before signing. (There was, however, an important mitigating circumstance, namely, Harry Hoyt's excellent reputation for dealing fairly with outside invention owners; at one time, he even went so far as to pay his R&D director a bonus for bringing successful products into Carter-Wallace from outside sources.)

Being Flexible

To reach a genuinely well conceived deal and to reduce that deal to paper requires creativity. You can read all the recommended clauses and procedures for drafting license agreements and you'll get a very good checklist of what to include. But whether you are an invention owner, a licensing officer, a lawyer, or a company president, please step back and ask yourself, "What is *different* in this case? What issues can we *foresee* and speak to now? What are the *substantive* points saying here? What points are *not* covered here that I may later regret omitting?" In short, different inventions raise different questions. Create answers to meet those issues and then plug the language into the license agreement.

For example, one of the most common conflicts in a license negotiation concerns what happens in the event of unlicensed competition. There are literally dozens of choices as to what shall happen if that occurs. Among the options:

- Terminate the license.
- Continue the license with reduced royalties or for a reduced period of time.
- Escrow part or all of the royalties until such competition ceases or is reduced below an agreed level.

Overlaid on any such action is the provision as to who is responsible for suing to stop infringement, and who bears the cost of suing, and who keeps the damages, if and when collected.

Now here's what I mean about being flexible. The simplistic approach to this area is to put the bee on the other guy, for the licensee to say to the licensor, "Heck, it's your patent and you're getting the royalties. You don't get a cent if there's unlicensed competition, and you have to agree to use your best efforts to stop such competition." Or the licensor may say, "You're a big giant. You've got the money. You've got the lawyers on staff. You're not paying me enough to afford big lawsuits."

The "Infringement" clause in my sample Heads of Agreement (at the end of this chapter) shows one way of handling this flexibly so that both parties' real interests, rather than their pride, are protected. There is no "right" answer in this infringement area. My erstwhile mentor in license contracts, Morton Amster, Esq., maintains that the parties should simply agree to work together in good faith to deal with any infringement that may arise because the economic realities keep changing as time goes by and ultimately people behave as their pocketbook dictates.

Being flexible does not include giving away your vital interests. For instance, if the agreement should be terminated, what happens to the licensee's inventory and work-in-progress? You certainly don't want it dumped on the market. Or what are your rights if the licensee doesn't pay the royalties due? There are soft clauses and hard clauses on this count; revoking the license is one of the best. All in all, pay attention to those clauses near the end of the agreement that tend to look like boilerplate and get language that flexes in your favor.

There's another form of flexibility that starts way back when you pick your prospective licensee. The obvious choice is not always the correct one. David James is CEO of Re-Tech, a company developing a revolutionary dry battery technology that is said to provide the output of a D-cell in a battery the size of a credit card. James approached people at Duracell, Inc., seeking a license or joint venture deal.

A top officer of Duracell reportedly told him: "I don't think you'll ever raise the money to scale this thing up. If you do and it works, I'll come to you and offer you $50 million. If you turn me down, I'll offer you $70 million—whatever it takes. Meanwhile, I've got a big investment in the kind of batteries we're making now." Not every businessman is so candid. In choosing your licensing target, you'll naturally focus on how much your prospect has to gain by licensing your invention, but it's equally important to study what he or she may have to lose or change.

A splendid example of choosing the right licensee was Dick Samuel's decision as to the first targets for licensing Gordon Gould's laser patents. Because the patents were still under reexamination and the laser manufacturers were still fighting them tooth and nail, Samuel's Patlex Corporation licensed major *users* of lasers such as automobile manufacturers. And these licenses included a provision that the company would pay a royalty of 6 percent on any future lasers it might purchase from nonlicensed vendors. Since the rate that Patlex proposed for licensed vendors would be only 5 percent, this created a future incentive for those laser manufacturers to sign up once the legal hassle was resolved. Had Samuel not been flexible in looking at licensing targets, years of royalty income would have been lost.

Whence Cometh My Help?

If you are smart enough or lucky enough, or a bit of both, to have the rights to an invention with the potential to bring you a million dollars or more from licensing, then you've already deduced from reading this chapter that it takes as much imagination, drive, and knowledge to sell the deal and get it properly buttoned up as it did to create the invention in the first place. Maybe more.

So you need allies. People who have the direct experience to help you with the selling, the negotiating, and the legal issues.

The problem is, whom can you trust? That's a tough question to answer specifically. For one thing, a lot depends on how well you read character. But I can tell you what seems to characterize the right people and, perhaps more important, what to avoid in the duds:

1. *Avoid being someone's first client.* If, for example, you're seeking a lawyer to write the license agreement, don't automatically

settle for your patent attorney unless he or she has substantial experience with licenses. Being good at writing and prosecuting patents does not necessarily qualify an attorney to negotiate and write license agreements.

2. *Don't pay consultants both cash and a percentage of the deal.* Whether you're taking on a lawyer or an advertising agency or a licensing service, the best arrangement is probably to have them work on a percentage basis. If you're convinced you have found people who can really help you, give them whatever percentage it takes to make them knock themselves out on your behalf. Remember, you don't take percentages to the bank. But if you can't make the right percentage deal, then pay cash but no percentage—and demand the best performance these people can deliver for that set amount of money.

3. *Be very careful about paying money in advance to anyone who claims he will do the entire job*—from patent search to patent filings to finding licensees, making the deal, and writing the agreement. Most companies that advertise on radio or in newspapers and magazines and claim to do all these things will do only one thing: take your money. Because I have some patents in my name, I get solicitations from these people by mail and by phone. My litmus test questions are: "Will you work on a straight percentage, with no cost to me whatsoever, getting your first dollar when I get mine?" and "How many inventions did you license last year, and will you please furnish me with the names and addresses of the inventors so I can check references?"

So how do you find good people? Whether you are a company seeking to outlicense or an independent invention owner, the answer lies in *networking.* Get out and meet people through their organizations: The Licensing Executives Society, The Product Development and Management Association, The American Management Association, The American Marketing Association, The American Intellectual Property Lawyers Association, and the professional societies for the industry at which you're aiming. Ask questions. Find out what new products or technologies have been successfully licensed in your general field and who were the key players, then enlist them on your team.

Inventor groups are of varying value. One group I know, the American Society of Inventors, in Philadelphia, provides a lot of help and encouragement for individual inventors through literature, seminars, and invention critiques. But you really have to go out and meet the group in your area and make your own judgments.

Chasing Rainbows, Chasing Money

Very few innovations are overnight successes. Sometimes the difference between the ones that make real money and those that make only a little is how seriously the innovator chases the gold.

My friend Walton Smith invented great things—like his semi-hydroponic planter that lets a plant water itself, drawing water as needed for weeks on end. It also makes it feasible to start growing tomatoes indoors when the ground is still frozen and to transplant the seedlings outdoors, pot and all, as soon as spring comes. Walt wrote the patent application and the advertising and invested in the molds and set up distributors and kept the inventory in his barn and shipped out the orders. People adored it. He got fan letters and had a lot of fun. And nobody bugged him. He was on the right path, but he didn't quite push it to the point where the world took enough notice.

Walt Smith stopped just short of what I call a kick-start start-up. That's when you start a company to sell enough product to show that there's life in the invention—and then sell or license it.

Tom Thornton did this and did it well. He really wanted to license his SuperFloss product; but while continuing his licensing efforts, he not only solved the manufacturing problems, he also built acceptance with dentists and founded a small business.

Earlier, I mentioned that Tom sold an option on SuperFloss to Pfizer, Inc., only to wait in vain for real action.

That wasn't Tom's only licensing effort. While still at the paper patent stage, he had had the product under option with Johnson & Johnson. Of that experience, Tom said, "I think they were actually mad at me for having invented it, typical N.I.H.* Anyway, they thought they owned the market and didn't need this product." Later, when Thornton had built the machinery to produce SuperFloss automatically, but still didn't have any sales, he negotiated a deal with Richardson-Merrill, Inc. That, too, fell through.

After seven years of fighting for strong patent claims, designing and building the machinery and opening small but profitable national distribution by promoting to dentists, Tom Thornton's company had all the earmarks of an actual business. Along came Parker Montgomery, then president of Cooper Laboratories, Inc., owner of Oral B toothbrushes. Montgomery, an astute merger and acquisition man, paid Thornton $3 million in an exchange of stock plus an

*Not Invented Here.

ongoing royalty on subsequent sales. How did Tom get a buy-out and additional royalties too?

In Tom Thornton's words:

> Parker Montgomery was both smart and fair in business. He understood where I was coming from in saying that this product could afford the royalty as just another cost of doing business. So the purchase price was the value they put on the going business, based upon profits generated on the existing business, net of the royalty they would be paying on such sales.

It was Tom Thornton's good luck that Cooper's Oral B division was sold off to Gillette, which beefed up the promotion behind SuperFloss, leading to a steady increase in royalties for its inventor.

If you should read all the books in my "Inventor's Bookshop" list, it will certainly improve your understanding of patenting and licensing. But you still won't know everything about patents and licenses and finance and negotiating. And life is short.

Successful license deals often rest upon having the right team of people, people with deep experience in each of the disciplines necessary for selling an invention and collecting millions. Not that reading books will hurt you if you concentrate on the areas where you are weak; at least you'll know the right questions. But buttress yourself with buddies who can match the technical, marketing, and business knowledge of the folks on the other side of the table.

Then, when the last compromises have been made, the final jots and tittles adjusted, the documents signed, the ink dried, and the documents properly stowed in the safe, please remember this: licenses are less permanent scriptures than, say, the Ten Commandments. When the deal is done, the work has just begun.

Start at once to create a partnership with your licensee by providing answers to technical, marketing, or business problems as they arise, or before. If the agreement needs to be changed to reflect what happens, take the initiative. With tangible products, that's called "service after the sale." With a license, it's called "continuity assurance."

Appendix A. Example of Heads of Agreement

Please note that everything in italics is for guidance; everything in roman is actual typical language. While this type of outline of terms covers the key points of the business negotiation, each case is differ ent and the Heads of Agreement should tackle all the major issues that you can anticipate. In addition, there are numerous points, such as the right to inspect records, how disagreements are to be settled, how trade secrets are to be protected, and others, that are not simply boilerplate; experienced legal counsel is indispensable.

Heads of Agreement Outline of Terms Used in an Exclusive License Agreement

Name of Product or Technology

1. Licensor:

   Your name or company name
   Address

2. Licensee:

   Customer's name
   Address

3. What Is Licensed:

   Describe the product or technology or application of the invention. If you want to limit the scope of the license, here is the place to start. For example, you can limit the product by weight, by size, by composition, etc.

4. Proprietary Rights:

   List patent numbers and patent applications with dates, country by country. Describe generally the know-how package.

5. Support:

   List what you will provide in addition to the proprietary rights. Examples: samples, ingredients, equipment, anything up to a turnkey operation.

6. The Grant:

   A [*type of license*]* license to make, have made, use, and sell the Product in the Territory under the Patent Rights.

   Also, a [*type of license*]* license to make, have made, use, and sell the Product under the know-how.

7. Option Period:

   State period of time, exclusive or nonexclusive, information and/or services to be supplied. Also option extensions, if any.

8. Option Price:

   State amount to be paid at signing or otherwise. Also, amount creditable against license payment or future royalties.

9. License Payment:

   State sum to be paid upon exercise of option or at signing if no option exists. State whether any or all of such sum is creditable against future royalties or otherwise refundable and on what basis.

10. Royalty Rate:

    Specify percent of net sales and periodicity of reporting and payments. Provide details if royalty rate changes by volume, by year, or other circumstances. Consider dividing royalty between patent rights and know-how so that the agreement continues if the patent should be lost.

11. Territory:

    State the geography.

12. Term of License:

    Concurrent with the life of the last-to-expire patent or a period of ten (10) years if no patent rights shall exist.

    While the foregoing is a good example, this is a negotiable item.

13. Minimum Royalties:

    An annual minimum royalty, payable in quarterly installments, shall be based upon approximately one-half of mutually agreed anticipated Net Sales.

    Minimum payments are 100 percent negotiable. The purpose is to ensure the licensee's full attention to maximizing the technology without being so punitive as to kill the deal. In most instances, a substantial advance royalty or minimum guaranteed annual royalty will help to ensure licensee performance.

**Fill in: Exclusive or nonexclusive or semiexclusive or sole. Note that patent rights need defining, for example, to include continuations-in-part, divisions, improvements. If you are licensing know-how, do so with a separate grant clause. If you are licensing a trademark or other rights, do so with a separate agreement.*

14. Term of Agreement:

    The Agreement may be terminated by either party by reason of an uncured breach upon sixty (60) days' notice. The Agreement may be terminated by the Licensee during the first five (5) years of the Agreement upon ninety (90) days' notice by paying to the Licensor the remaining minimums for the said first five (5) years of the Agreement. After said first five (5) years, Licensee may terminate the Agreement at any time without penalty upon ninety (90) days' notice.

    The foregoing is one way to help ensure a Licensee's best efforts at the outset. It is always a negotiable area. As a guideline, consider termination with a single payment equal to the minimums for the succeeding five years or so of the agreement. Be sure to provide for disposition of unsold inventory.

15. Patent Filings and Maintenance:

    Licensor shall be responsible for all patent filings, for prosecution and maintenance of patents, and for all costs relating to same.

    Keep control of your own patent work. If you need financial help, ask the Licensee to reimburse you for any filings made at his behest, for example, foreign filings.

16. Improvements:

    Best case: Include without cost all improvements made by the Licensor and get back a nonexclusive right to use and license outside of the territory any improvements made by the Licensee.

17. Indemnification:

    Licensee shall hold Licensor harmless from any and all product liability claims caused by manufacture and/or sale of the Products by Licensee, excepting for direct negligence on the part of the Licensor.

    Licensee has insurance; licensor doesn't and can't afford it. A "must."

18. Infringement:

    Licensor shall have the first right to bring appropriate action against infringers at its own expense and shall retain any recovery therefrom. If Licensee shall fail to take action within ninety (90) days after learning of such infringement, then Licensor may institute appropriate action at its expense and retain any recovery therefrom.

There are an infinite number of ways to handle the issue of infringement. The above is reasonably balanced, but the situations vary widely. Issues include possible escrow of royalties, reimbursement of costs, etc. The best solution often is for Licensor and Licensee to simply agree to work in good faith to deal with any infringement situation.

IMPORTANT NOTE: THE FOREGOING OUTLINE OF TERMS IS PROVIDED FOR INFORMATION PURPOSES ONLY AND DOES NOT CONSTITUTE AN OPTION OR AN OFFER TO BUY OR SELL.

Chapter 12

Start-ups: The "Hard Way" to Money

In the previous chapter, I labeled licensing the "easy way" to turn an invention into serious money. As you read that chapter, you may well have thought, "Easy? I don't see anything easy about licensing in the examples you've given."

Of course, everything's relative. Instead of calling it the "easy way," perhaps I should have characterized licensing as the "direct way" or the "quickest way." Nevertheless, compared to starting a company from scratch, betting on a licensee is easier than a start-up, especially as measured by strain and pain.

On the other hand, although starting your own company may be the "hard way," it has some key advantages. You can keep control of your innovation's destiny; you can make adjustments when technical or marketing problems arise. You also have a business vehicle for additional inventions. And the odds are that you will make far more money this way than you would by licensing, especially if you build your young company to a point where it can be taken public or merged with or sold to another company.

Starting a business to exploit an invention is different from, say, opening up a Pizza Hut franchise, where operating pathways are defined. A new-new product is a paradoxical animal: The more fantastic and revolutionary the product is, the fewer the known pathways to follow and the greater the need for business creativity. Financing, manufacturing, and marketing are all impacted by this dilemma.

When you go to raise capital for a new product business, investors must make a leap of faith to believe that the product can be manufactured, that it will sell, that it will make a profit—in sum, that you can operate the company as your business plan predicts. The more genuinely new the product, the harder it is to prove viability in advance. It requires vision, and vision, as Jonathan Swift once wrote, is "the art of seeing things invisible," a talent not commonly found in the financial community.

Of course, there's a blurry line between seeing things invisible and seeing things that simply aren't there; the major problem of an inventor/entrepreneur is to convince investors that, as was said in *Field of Dreams*, "if you build it, they will come."

As every experienced businessperson knows, young businesses, like children, need not only to be protected but directed. Protection for an invention-based start-up business is found in patents and other proprietary picket fences—and money in the bank—all of which buy time to grow. In addition, directing such a young business and raising the brainchild requires as much imagination and innovation as the invention itself.

Clearly, not every invention-based start-up makes a fortune for its founders. So, what are the hallmarks of the winners? Can inventive people carry their creativity over to the business side? Can they find and get along with unusual allies to solve the inevitable start-up problems?

As you will see from the many stories of successful start-ups in this chapter and the next, while the specifics differ, the inventors and entrepreneurs who make millions from the mind have a lot in common.

Inventing Answers to Business Problems

A well-known characteristic of successful entrepreneurs is that they understand the word *no* to mean "Charge!"

David Montague started his company to market a bicycle that folds so that it can be easily transported or stored. Unlike most folding bicycles that have small wheels and spindly frames, the Montague BiFrame is a full-size bike with 18-speed index shifting, cantilever brakes, a chrome-moly frame—the whole works for a "mountain bike." The ingenuity that permits a full-size bike to fold lies in the vertical double tube on which the seat rests: You release a couple of latches and the rear half of the bike folds back on the front end. At present, it retails for just under $500, comparable to the price for good-quality conventional mountain bikes.

The Montague bike was invented by David's father, Harry, a full-time architect and part-time inventor in Washington, D.C. Originally, when he got his first patent, Harry Montague hoped to license his invention, collect royalties, and peacefully practice architecture. So he dashed off a proposal with a blueprint to most of the leading bicycle manufacturers in the United States and abroad. His offer had two results: a series of polite letters, all saying "no, thanks" and, about a year later, the appearance of a Chinese copy made by a Japanese company, the subsidiary of a very big Japanese conglomerate. Fortunately for Montague, his patent lawyer, Neil Siegel of the Washington firm of Sughrue Mion Zinn Macpeak & Seas, had moved the patent through the Patent Office deftly, leaving a perfectly "clean" file. Siegel helped the Montagues convince the Japanese that, while they would fight to the end, an agreement with honor for both parties was possible. As a result, the Japanese settled out of court, agreeing to pay a royalty on what they had manufactured and agreeing not to sell in the United States. The value of such a settlement, as a message to other would-be infringers, cannot be overestimated, especially for a start-up company.

Harry Montague originally launched the product by having fifty units made, using his original model, a racing bike. He sold them, catch as catch can, for $1,400 each. Son David, while still at MIT studying business and aeronautics concurrently, began working on the bicycle project. David along with his classmate John Nelson formed the Montague Corporation to market the BiFrame bike in earnest. One of their first moves, timed perfectly to meet the changing public taste, was to move from a racing bike to a mountain bike.

Early on, Montague Corporation faced several hurdles common to start-ups. There was no manufacturing plant, no distribution system, and only limited capital from family and friends (ultimately, $300,000). On the plus side, however, there was youth, great imagination, and the willingness to do whatever was needed to win.

Bicycles are made all over the world, so David did his homework, checking quality, size, prices, and so forth through foreign commercial attachés and trade associations and published sources. He found out that 80 percent of the bicycles sold in the United States are made in Taiwan. So he did the "absolutely indispensable" thing. He got on a plane. He journeyed from plant to plant, met the people, saw what they produced. His reception was mostly chilly. No one wanted to do business with an unknown young man lacking an established company.

But as they say in baseball, it takes only one, and David found his one in Taiwan, a maker who agreed to accept his order, provided it was accompanied by an irrevocable letter of credit.

As David recalled to me, "In a new entrepreneurial business, *nothing is for sure*. And when you're doing business in the Far East, if you don't get out there and meet those people and find out firsthand what's going on, you will fail in any import business." Wisdom from a twenty-six-year-old!

To market the Montague bike, David decided that he needed a distribution system that would provide product service for the consumer. That ruled out catalogs, direct sales, mass merchandisers, and department stores. He decided on mainstream bike specialty stores reached via distributors.

Initially, David visited a few bike specialty stores to probe dealer reaction. He talked with bike owners too. He could have taken orders on the spot, but he wasn't ready to deliver.

Then he hit upon an arrangement that would solve his distribution and financing problems simultaneously. He sought exclusive distributors for major territories and convinced these distributors that they should pay for their orders by a letter of credit to Montague Corporation's bank. Based upon that LC, he got his own bank to issue a letter of credit to the Taiwanese.

Now, is that inventive—or what?

Persuading people with excellent credit ratings to pay for inventory in advance of delivery from a start-up company is what my grandmother would call "one neat trick." When I asked David Montague how he was able to persuade distributors to do this, he said, "Well, we had a totally unique product and a completely finished product. When we showed the BiFrame, it worked perfectly. We had our production source, our promotion material, everything was set. So it was a clear profit opportunity."

But it was only after exploring deals with venture capitalist people who wanted half of his company for financing it that David was forced to think of his ingenious alternative.

At the outset, Montague used regional distributors, but he recognized that these people left big gaps in the retail distribution pattern. In addition, it became apparent that the BiFrame could not be sold by "push" through bike dealers: The salespeople in those stores were mostly young, often transient, and there was no way to train them to sell a product they didn't understand.

Montague moved to solve the latter problem through limited advertising and heavy publicity that drove customers into the stores. Then, through a friendly intermediary, he sat down and cut a deal with the Schwinn Corporation to become Montague's national distributor for the United States. Schwinn, with forty-five salesmen, warehouses throughout the country, and an established reputation as the number one U.S. bicycle manufacturer, game Montague both prestige and distribution. The product became the Schwinn Montague BiFrame.

Montague continued to source the product and conduct the marketing activities; Schwinn warehoused and sold to the dealers.

The Montague strategy for getting the bikes out of stores and into the hands of consumers called for advertising only to the obvious "prime prospect" markets while creating a whirlwind of broad media publicity. Early on, David recognized that he had to reach four key target groups: (1) serious biker riders, especially in metropolitan areas, where security for the bike and convenience in transporting it are of real concern; (2) private plane pilots, who often find themselves at a local airport without transportation; (3) boat owners who need shore transportation and have little stowage space for a regular bike; and (4) recreation vehicle owners who have similar storage problems and don't want their bikes "hanging out."

Advertising in special-interest magazines covered these primary target groups even as publicity, generated mostly by an internal PR person, generated more than 350 press and TV exposures in less than two years. Having such a demonstrable invention was a key to such strong publicity results.

One major communications problem was how to distinguish the Montague full-size bicycle from previous slow-moving folding cycles with little wheels. David's answer was summed up in a slogan: "The BiFrame is a bicycle that folds . . . not a folding bicycle."

So the Montague product was terrific. The financing method was clever. The distribution network generated orders all over the country. The publicity and advertising were highly effective. Under the law of probability for start-ups, something was due to go wrong. And it did. David explains:

> Our first big blow came when our manufacturer in Taiwan was four to five months late with our first delivery. It seems they underestimated the time for tooling up. This cost us a whole season because the bike business—big bikes, not toy bikes—is seasonal, April through the end of the summer. We missed one whole season.
>
> Another crisis developed when we received the first three hundred bikes with the Montague decal upside down. We stood there for days and patiently peeled them off and applied new ones.

Aside from such commonplace aggravations, Montague is the story of a parlay of inventive "right moves," a pattern that many inventor/entrepreneurs might follow. David Montague and John Nelson began in the summer of 1987. By the spring of 1988, the company was incorporated and by June the first shipments began. By the end of 1990, the company was going at a rate of $4 million a year in sales, highly profitable, and growing and exporting too.

Starting Up as a Road to Licensing

If you wanted to nominate invention categories for the least-likely-to-produce-riches award, you'd have to put concrete building blocks pretty high on your list.

For years, people considered efforts at innovation in the building block field akin to running into a stone wall. There are many fields like that, industries in which there has been little inventiveness and much stubborn resistance to change. Most people find it difficult to license inventions in such fields, and manufacturers of concrete blocks may well take top billing.

But Robert Dean, president of Designer Blocks, Inc., of Milwaukee, Wisconsin, isn't "most people." He found the answer by first building a business using the technology and then licensing both the invention and the business know-how.

In 1980, while working as sales manager of Best Block Company, he yearned to put into practice the theory of product differentiation that he had learned while getting his M.B.A. at the University of Wisconsin. One day, as he stared at a cut-stone wall, he asked himself why concrete blocks had to be flat: Why couldn't they be cut in varying planes to create a random pattern, a cut-stone texture?

The more Dean studied the idea, the more he believed in it. In a relatively short time, Dean invented a way to set the automatic splitting machine that cuts concrete strips into blocks so that the depth of the blocks would vary by predetermined amounts. A patent search showed no prior art at all so a patent application was filed and in due course issued. Designer Blocks® were a reality—on paper.

As expected, he encountered rock-hard resistance to change. Dean pointed out to me, "Getting the idea and the invention and the product wasn't nearly as hard as the marketing."

Then, with foresight rarely exhibited by employee-inventors, Robert Dean persuaded the owners of Best Block to set up a separate company for Designer Blocks in which Best Block would have stock, but which would be run and controlled by Dean, a deal that proved profitable for both parties. Please note that this is the kind of "intrapreneurship" that works best—because it started with a strong, defined, and proved product and, to give it a separate identity, was split off completely from the parent, excepting for the stock ownership.

The first year, Dean showed the product at an architects' convention and received promises of serious consideration for new buildings. He also set up a display panel of Designer Blocks at his offices, to which he brought key people from every side of the building trade. In the initial year, Designer Blocks were used for twelve buildings, a remarkable accomplishment in an industry that does not quickly embrace new ideas. Why did it work out so well? Dean's salesmanship counted heavily, but the visible impact of the product and a competitive price versus conventional blocks were the critical factors.

Then Dean faced the real question. He had a nice small business in his hometown area; but how could he expand the business?

There are only so many new buildings going up per year in the Milwaukee market area, and only a small percentage of the new structures were candidates for Designer Blocks. Because concrete blocks are extremely heavy, it was not economic to ship them more than, say, 150 miles. How could he develop Designer Blocks into a national business if he couldn't ship the product?

The obvious answer was a franchising/licensing arrangement. But what was obvious wasn't simple. As Bob Dean explained it to me,

> When we went to local concrete block manufacturers, we didn't offer just a license. We set up a full marketing support program with a complete manual on how to promote and sell Designer Blocks. We provided brochures for every kind of potential customer and influence group. We even had a slide presentation showing various applications of Designer Blocks. We offered to do mailings to architects with the local licensee's imprint if the licensee provided a list of names.
>
> And what we learned was that we still had to do a lot of hand holding. Even though Designer Blocks were more profitable for the local block manufacturer than standard concrete blocks, most of them were organized only to negotiate price, produce, and deliver orders, not to promote and merchandise, not to *market* a differentiated product. As a result, 25 percent of our licensees do a real job while the other 75 percent make some money but don't realize their potential.

There are two ways to attack such a marketing dilemma. Either you set up a specialty sales force in the field to "live with" your franchisees until you get them up to speed or you look for additional product opportunities in your field of expertise. Dean made a modest attempt at the former route, but, with a comfortable $200,000 a year flowing in from Designer Block royalties, he allowed a second invention to draw his efforts in another direction.

Why not, Dean reasoned, look at other kinds of walls in which stone or wood is now used and invent a concrete block replacement? A study of possibilities quickly focused his attention on retaining walls, the kind used for maintaining ground levels and controlling runoff. Dean analyzed the competition: wood, which required considerable work to install and was subject to deterioration, and stone, which required heavy, skilled labor to set.

As with many good inventions, the answer came to Dean after the problem had percolated in his mind for a while. He conceived of notched concrete blocks that could be put together as easily as a child builds a wall of Legos. In short order, he founded Stonewall Landscapes, Inc., and by mid-1990 had licensed fourteen franchisees.

This time, however, mindful of his experience with Designer Blocks, rather than trying to turn concrete block peddlers into promotional marketeers, Bob Dean found experienced marketing people as his area franchisees and then teamed them up with local concrete block makers. In addition, he built in an economic incentive for each party, something that was possible because the Landscape Blocks—unlike the concrete blocks—could be priced to allow adequate margins for marketing expense.

Pricing the Landscape Blocks to their value in the market also had a dramatic effect on his royalty: forty-five cents per block for the new Landscape Blocks versus five cents per block for the Designer Blocks. Dean expects the royalty stream from Stonewall Landscapes to grow three or four times as fast as that for Designer Blocks because he has created a "big company structure" by the ingenious combination of his own central marketing and promotion group, regional franchise owners, and local block manufacturers.

Making a Business of "Small" Inventions

One of the very brightest men in the world of invention is Rick Onanian of Arlington, Massachusetts, near Boston. He is an inventor of such unusual items as the Super-Microscope and the Ultra-Microscope, scopes with remarkable resolution that sell for only $18 and $12.50 respectively, providing schoolchildren with an exciting window to science. He is also a person with experience in licensing and in financing ventures.

He now owns his own business, a company that sells educational products via catalog to schools across the United States. It was a twist of fate that turned him from a classic inventor into an inventor/entrepreneur.

Rick had invented and licensed several educational toys to Learning Things, Inc. When the owner of that company passed away, successor management fell in arrears on Onanian's royalties. Faced with the prospect of losing not only the back royalties but the entire market that had been developed for his products, Rick talked with the bank that held the key indebtedness of the company. He promptly assessed the situation. Although he had no power to collect his royalties, the bank couldn't sell the company to anyone else because his license agreement called for the rights to revert to him in the event of bankruptcy or sale of the company.

That reversion clause is frequently found in well-drawn license agreements because no one wants their patent rights to be tied up with a company in the bankruptcy courts. But for Rick Onanian, reclaiming the patent rights simply meant starting again on the long trail to finding a licensee—and for a no-longer-new product at that.

So instead, he made a deal with the bank to take over the entire company on a "fire sale" basis.

Rick's experiences with licensing primed him for that deal. Several times, inventions had come bouncing back to him after short option or license periods. He was, in fact, a bit bitter. "There are too many incompetents masquerading as executives in large corporations today," he told me, "to put your invention in their hands. If a new product development comes out of a company's own lab, it receives the benefit of all the company's resources, but when a company goes outside and licenses a product, it doesn't have the same emotional attachment to it."

Whether or not that indictment is fair, it is certainly the view held by many independent inventors and not a few mainstream corporate executives. In any event, it motivated Rick Onanian. He built Learning Things, Inc., from a line of 10 products to a range of more than 400 items. Ten percent of them are his own inventions. Generating most of his business from catalog sales to schools, Rick Onanian created not only a profitable niche business but a continuing marketplace for his own inventions. And without royalty collection problems.

Direct-to-the user selling is a favorite channel for young companies launching specialty inventions. Strength Footwear, Inc., of Metairie,

Louisiana, markets its Strength Shoes via mail order to school teams and to individuals. The Strength Shoe is a strange-looking athletic shoe with a 1½-inch-high rubber platform under the forward half of the shoe. Marketing VP Rich Sheubrooks says that it builds up power and quickness and the ability to get airborne. Using only publicity, traveling demonstrations, and limited advertising in such magazines as *Sports Illustrated*, Strength Footwear has sold several million dollars' worth at $99.50 a pair. Protected by a patent licensed from the inventor, Paul Cox, a physicist, Strength Footwear is an example of how direct selling can turn a "small" invention into a viable business.

The Big-Time Small Start-up

It's true that a start-up company can be set up and run with the sophistication of professional business management—provided you have three things: people with big-time experience but with entrepreneurial souls *and* a unique product.

Even with those qualities, to springboard a product in a category that has been considered moribund for years is quite an assignment.

That was the case with Chef's Choice, the electric knife sharpener introduced by Edgecraft Corporation in 1986. As had happened with Hula-Hoops, before they were introduced, if you just told people about this idea, they yawned. Before Chef's Choice, there was no hue and cry from the kitchens of America for a new and better knife sharpener—certainly not one that cost the better part of a hundred-dollar bill. Nor would a survey of housewares departments have revealed that store buyers had any interest.

Enter Daniel Friel, inventor of Chef's Choice. He spent a long and illustrious career at Du Pont, where he was put in charge of a series of start-up businesses based on scientific developments. In 1980, when he decided to work for Du Pont just three more years and then retire, he began to spend every moment of his free time seeking a way to make a better knife sharpener. An odd pursuit? Not to Dan Friel. He loved things that worked well, he knew that the knives in his own house never had a proper edge, and he checked all the sharpeners on the market and found them wanting.

In his own mind, Dan drew a parallel between his super-sharpener idea and Cuisinart food processors: He saw two relatively expensive products, each catering to an unfulfilled need, each performing dramatically better than anything previously on the market. He was confident that many people would cheer him with their pocketbooks but only if he could produce a superb product. He set a standard for his sharpener: "twice as sharp as a razor." That phrase told everyone involved in the engineering that "good enough" just wasn't good enough to create that invaluable momentum which comes from word-of-mouth recommendation.

By 1983, two years after he left Du Pont, Dan Friel had produced working prototypes. By 1985, he had incorporated as Edgecraft Corporation and built enough samples to do consumer testing and make his initial sales.

Dan's product is an electric-powered knife sharpener with three slots for grinding. The first slot roughly shapes the blade edge. The second slot hones the burrs from the rough shape. And the third slot burnishes the cutting edge. As you move the knife through each slot, magnets hold the knife at exactly the right angle relative to the grinding wheel.

While Chef's Choice is an engineering triumph, it turned out to be an achievement in market research and marketing too.

With working models in hand, Dan Friel started with focus groups of people selected because they not only had homes and families and kitchens but because they could afford "luxury" items as well. As Dan described the process to me,

> We tried to establish a baseline by asking them what they thought a really good knife sharpener should sell for, how much they themselves would pay for one if it existed. Some said ten dollars, some twenty—I think the highest was forty dollars. Then we asked a lot of others questions about their spending habits and values to see if those answers were sincere or just floating replies.
>
> Then we took out our prototype sharpeners and asked these same individuals to sharpen a knife and slice a tomato with it. We asked them to describe the performance and, most important, to tell us how much they would pay for such a sharpener. The amazing thing was the consistency of the answers: They told us they would pay twice as much now that they had seen the product perform. If they had said ten dollars before, they said twenty now. Twenty went to forty, forty to eighty.
>
> Since we knew the income levels of our panelists, we sorted out what percentage of them had said they would buy at the eighty-dollar level and applied that to the national income distribution numbers to find how many people might be in our target market. Of course, that was a

> rough cut, but it amounted to 10 percent of the population. We had several hundred people interviewed, statistically perhaps not enough for extrapolating sales forecasts, but enough that we knew we had a business.

Friel had, in fact, engineered the product so that it could be sold at retail for $50. But that would have meant minimum margins and limited promotion money. So toward the end of 1985, Edgecraft built 500 Chef's Choice Diamond Hone Sharpeners and made placements at Macy's, Jordan Marsh, Cutlery World, and several different chains around the country, pricing the product to retail at different prices in different places—$59, $69, $79, and $89.

"The result," Dan Friel told me, "was surprising. There was very little difference in sales between a $69 price and a $79 price. So we went national at $79. You know, you can always come down."

Dan Friel vastly admired the job that Carl Sundheimer had done in launching Cuisinart food processors, and he followed that pattern through the upscale stores, avoiding the discount stores as best he could.

Edgecraft was able to go national very quickly because Chef's Choice offered retailers a new source of profit. It was not an item that would cut into their existing sales of other kitchen equipment, except to the extent that everything competes with everything else for the consumer's dollar. With successful test marketing in a few stores to prove turnover, Edgecraft rapidly built a sales force by tying in with manufacturer's reps who handled top-quality lines like Calphalon and Krups, a sales force that was able to sell almost all of the prime outlets, including Bloomingdale's, Hammacher-Schlemmer, and Williams-Sonoma.

But Friel went several steps further, and it was a marketing coup. First, he sold to the major catalog firms such as the Sharper Image simultaneously with obtaining department store distribution. This not only gave Edgecraft additional volume, it provided high product visibility, advertising that, instead of costing money, created profits.

Next, he turned to PR. Dan Friel showed remarkable perspicacity in stretching promotion dollars. His experience at Du Pont working with many-zeroed budgets had taught him the value of PR: "If I had a budget of $5 million a year and put $4.9 million of that in advertising and $100,000 in PR, most of the time I found that the PR gave me more benefit."

Using his own people to do the spade work, Friel was able to secure favorable press stories from Pierre Franey, the respected culinary equipment reviewer for *The New York Times,* and many other writers around the country. Craig Claiborne, dean of the gourmet food reviewers and a great chef in his own right, became enamored of the sharp edges produced by Chef's Choice and agreed to appear in Edgecraft's small ads in *Gourmet*, the *New Yorker*, and other upscale magazines.

To accomplish this unusual success story Friel required a product that was not only new but newsworthy, that not only delivered on its promises but gave a truly superior performance. And, I must add, had a patent position to provide the running time needed to cream the market and get on to the next product. Edgecraft had four issued patents and one pending as of early 1990. The patents cover both the magnetic guide system and the orbital grinders, the heart and soul of why this invention allows any dummy to put very sharp edges on kitchen knives.

The follow-up product from Edgecraft Corporation turned out to be a scissors sharpener. At first blush, this doesn't sound as exciting as the knife sharpener, but listen to Dan's rationale:

> A number of people in the trade and a few consumers suggested a scissors sharpener to us. We knew the market was narrower, but when we looked into it, we found that there are 23 million sewers in this country, about half of them what you'd call "serious sewers." It also turned out that when you want a pair of scissors sharpened, you can wait up to three weeks to get them back. People who use scissors in commercial businesses for cutting cloth or cardboard consider this a real problem. Then, too, our technology and patents from the knife sharpener are directly applicable, so we could make these units without major R&D.

That man thinks before he leaps.

Dan Friel and one outside investor have put well over a million dollars of private money into starting Edgecraft Corporation. The company moved into the black three months after Chef's Choice was launched. After two years of operations, profits were so satisfactory that Dan Friel began looking for compatible acquisitions.

Turning One Product Into a Major Business

Many an inventor would prefer an all-cash deal to sell the invention early on, to "take the money and run," so to speak. But because it's hard to tell what a newborn invention is really worth, this seldom happens.

Here is a neat example of how an invention grows in value when you hang in there and preempt the market before the world catches on. It is the story of Betty Nesmith's Liquid Paper.

Prior to the invention of self-correcting ribbons and computers, there was no satisfactory answer to the basic problem of every typist: how to correct mistakes on the original and still produce a professional-looking document. For years, most secretaries used a rather imperfect invention, a round, hard typewriter eraser with a long-bristled brush attached to flick away the granules that flaked off the eraser. It eradicated the error all right, but often took the paper with it and left an unsightly hole.

In Dallas in 1951, Betty Nesmith was an executive secretary to the chairman of a bank and, by her own admission, a poor typist. And her boss knew it and was growing testy. She was also a divorcée with a son to raise, and she needed her job. Her new IBM electric typewriter produced sharp images but left the page smeary when she erased an error, and Betty made her share.

One night, she invented what became Liquid Paper. Using tempera water-based paint—something she'd seen artists use to correct lettering mistakes—she produced a viscous liquid that could be dabbed on a misbegotten letter or word, opaquing the error and providing a new surface on which the correct letter(s) could be typed.

Betty Nesmith nursed the invention from a kitchen-production, pin-money business to an international company selling more than 25 million bottles of Liquid Paper annually. In 1979, Gillette bought her out for $47 million plus kickers.

Start-up as a Prelude to Sell-out

In the real world of corporation executives—that world out there where paper profits are the quarterly prerequisite to career advancement—fear of failure with new products often paralyzes executive thinking.

This explains why big corporations would rather pay a big multiple of sales for a start-up company—even if it's losing money hand over fist—than license a great invention that has not yet reached the market.

That was certainly the experience of Ron Hickman with his WorkMate®. The WorkMate, for those of you unacquainted with the do-it-yourself world, is a small, sturdy, highly portable folding work surface that is rugged enough for hammering, sawing, drilling, or what have you.

Hickman, an Englishman with bulldog determination, took the invention to most of the logical major manufacturers in England and, reportedly, made good presentations to senior management in such companies as Black & Decker, Stanley, and other known factors in the tool business. The universal verdict was that the WorkMate didn't have any large market potential.

In the face of this universal disdain, Ron Hickman turned to his own resources. He began to make and sell the WorkMate himself. He contracted the parts manufacture, did a "garage assembly" job, and sold the product to a few stores and by mail order. Within four years, annual sales reached 14,000 units. He kept improving the product based upon user feedback, adding a swivel for the vise beams, height adjustment, and so forth.

At this point, the general manager of Black & Decker in the United Kingdom approached *him*. Hickman sold the product rights to Black & Decker in 1972, "selling out" via a license agreement since he had no tangible assets. The product became an immediate smash not only in Great Britain but all over Europe. It was one of those products with instant impact for the do-it-yourselfers—even though people as experienced as Black & Decker had not foreseen this when it was offered to them in prototype form, possibly because WorkMate created a whole new product category.

Eventually, Hickman signed three licenses with B&D, the initial one covering the United Kingdom, then Europe, then North and South America.

An interesting feature of this story is that Hickman's royalty rate was only 3 percent. This was considerably less than his counsel, the very experienced British patent lawyer M. J. Roos, thought he was entitled to. But Hickman felt that, by accepting the 3 percent rate, he moved the product into strong hands early on, so that when the imitators came along there was enough cash flow to justify the cost of litigation, and the patent was very successfully defended by Mr. Roos.

In terms of Ron Hickman's rewards from his invention, although he has certainly prospered, there is a bit of cautionary history in his license agreements. The ones for the United Kingdom and Europe provided that Black & Decker—an American company, remember! —could buy him out for a flat sum. The agreement for the United States, by far the largest market, contained no such clause and B&D was obligated to continue to pay royalties for the life of the patents.

As recounted in Chapter 11, another example of the build-to-sell principle was Tom Thornton's seven years of persuading the dental profession that SuperFloss was an interproximal brush and not just plain dental floss. No company would license his product, but when that base of professional acceptance was achieved together with a few million dollars a year in sales, he sold his small operating company to Cooper Laboratories in what he described to me as, "an easy, pleasant negotiation."

As Hickman's and Thornton's and Nesmith's stories clearly illustrate, whereas many corporation executives are afraid to buy or license an invention, they are less reluctant to pay out much more money if the transaction is called an "acquisition." That's probably because acquisitions are approved by the board of directors—and who ever fires them?

Not Just a Promise, but a Threat

While this book is primarily about American inventors, I included the Englishman Ron Hickman because his product ended up with an American company. Now I want to tell you about Ralph Sarich, an Australian, who invented the Orbital Combustion Process gasoline engine, an innovation that led him to manufacturing in the United States.

Besides, from an entrepreneurial viewpoint, Australians are just Americans a hundred years removed, aren't they?

As this is written, both Ford and General Motors, among others, have taken full licenses for Sarich's Orbital engine. Events in the Middle East have put the spotlight on the Orbital engine's economy. And Sarich has placed his first-ever motor production plant in the United States at Tecumseh, Michigan (aided by an $80-million support package from the state of Michigan), and has a joint venture with Walbro Corporation to produce his fuel injection system at Cass, Michigan.

You could hardly find a more difficult field in which to make your fortune than creating inventions to sell or license to the automobile companies. GM alone gets 30,000 to 40,000 "ideas" submitted annually. As thousands of inventors who have been rejected out of hand in Detroit will testify, getting any of the auto companies even to examine an invention is a real challenge.

Sarich, based on his experience and training, would hardly have seemed destined to break the rules. A farm boy in Australia, the son of Yugoslav immigrant parents, he was without formal training as an engineer, a former fitter-and-turner who in 1970 was selling earth-moving equipment when he quit his job to pursue a vision of an engine small and light enough to serve as a portable power plant on a farm.

But he had two unique qualities: He could invent solutions in his mind that solved "impossible" problems conceptually, and he was and is a superb, low-key charismatic character who will not take no for an answer.

Sarich's first attempt at revolutionizing automobile engines almost ruined him and left behind traces of doubt that haunted him. Modestly supported by two investor-partners and by doing contract engineering work, by the summer of 1972 he had developed a very compact, light-weight, four-cycle engine.

Unwittingly, he became a hero of the media. Perhaps it was the romance of finding someone in a small shop near Perth in Western Australia, perhaps it was Sarich's flair for dropping remarks about the potential performance of the next generation of his engine, perhaps it was the 1972 Mid-East oil crisis—whatever—but the press took up the Sarich story and made him Australia's new savior. Automobile trade publications praised him and the Australian Broadcasting Company featured him on network television in a show called *The Inventor*. Without regard to the relatively minimal value of his first engine development, Sarich became a fabled figure in the gung ho business atmosphere of Western Australia.

Early on, Sarich had invented some nonautomotive products—a monster off-loader system and a unique sprinkling system. He licensed them to keep afloat and this provided some credibility. Of major importance, however, was the fact that in 1972 Broken Hill Properties (BHP), one of the largest conglomerates in Australia, always interested in diversification from its original mining business, supported Sarich's developments with $1.2 million, a sum that was eventually increased to some $20 million. Then, the state of Western Australia provided supporting grants and eventually a site for a new R&D facility.

During the early 1970s, Sarich's next development, a rotary "Orbital" engine, was hyped by the press and by financial people but never went into production. As a result, it was successively praised, then questioned, and finally scorned. The Wankel rotary engine had seized the initiative with car manufacturers.

At this low point in his career, Sarich turned his attention to a four-stroke engine using his "stratified charge" invention to create a new ultrafine fuel injection. The effort paid off in a prototype that was very small and efficient by comparison with conventional four-stroke engines.

By 1983, just when General Motors was at last ready to sign a contract with him for substantial money, Sarich changed his mind. He decided that it was short-range thinking to dedicate the stratified charge invention to a conventional four-cycle motor when he could use it to make a truly revolutionary engine, a two-stroke engine that would be cheaper, smaller, and lighter with amazing fuel economy and reduced emissions. Very simply put, he foresaw converting a simple two-stroke engine into a smooth, powerful, economical power plant. Bear in mind that two-stroke engines take in air and fuel and expel the combusted gases with every round trip of the piston. They sound like a lawn mower put-put and spew out pollutants. Four-stroke power plants are smoother and quieter.

Sarich had bet right. The Orbital combustion system used a two-stream fuel injector and the stratified charge principle to produce a two-stroke engine with startling characteristics. Some of its advantages include 50 percent less weight and 70 percent less volume than conventional engines of comparable horsepower. In addition, gasoline mileage was improved by 20 to 25 percent for automobiles and by about 40 percent for marine engines. And the Orbital engine required about 250 fewer parts than current engines do and cost about 25 percent less to manufacture.

Who says all this? Sarich Technologies Limited says it, of course. But so do half a dozen automobile review magazines in both Australia and the United States. And most significantly, General Motors and Ford and Outboard Marine and Mercury Motors have said it by paying Sarich Technologies millions of dollars up front. No one aside from the contracting parties knows the terms for certain, but the consensus of undenied analyses is that the licensees will be paying about $30 per engine in royalties.

Why? Why would big corporations pay out millions of dollars for licenses to an innovation in a field where they have great expertise and pride? The answer is that the Orbital engine was not just a series of promises of better performance but a threat to the market share of any auto company that didn't have access to it. It's a powerful truth in licensing that fear moves people at least as much as greed. Fear of missing out, if the innovation is reasonably documented, can motivate even the hardiest skeptics.

Sarich Technologies Limited went public in 1983, underwritten by two respected Australian houses, Hartley, Poynton & Company of Perth and Potter Partners of Melbourne. It made BHP's investment look good and it made Ralph Sarich and his family among the wealthiest people in Australia. Within the first year, the market valued the company at $930 million, quadrupling the price per share over the original issue price. In the years thereafter, the stock price swung wildly up and down in reaction to both the general market for new technology stocks and the continual announcements of progress by the company followed by periods of waiting, waiting, waiting for at least one automaker to announce that an Orbital engine would be used in a definite model and year.

But as the world press continues to report that the Orbital-Walbro fuel injector is ready for marketing, that the engine plant at Tecumseh, Michigan, is moving rapidly toward completion, that Ford and GM as well as Bajaj Auto (the largest motorcycle maker in Australia) all plan to launch an Orbital-powered vehicle in 1993, the credibility of Sarich Technologies has been sustained over an incredibly long period of time, some twenty-two years.

With more than 420 patents and applications in twenty-one countries around the world and six full licensees, Sarich can project royalties and profits from his inventions that are positively mind-boggling. They would, if realized, make him one of the richest men in the world. And even if the Orbital engine should fail, no one could call Ralph Sarich a failure.

How did Sarich do it? What distinguished him from untold thousands of inventors in their garages who have modified, rectified, rarefied, specified, magnified, dulcified, and glorified automobile and motorcycle engines?

First, he is a rock-solid pragmatist, not a theorist. He wants to know, does it *work*, and if it doesn't work, *why*, and what are we going to do about it *now*? He is absolutely focused on the problem at hand. He knows no hours and expects the same of people around him.

Second, he has a gift for self-promotion under a modest exterior, a very special combination. The difference between effective self-promotion and braggadocio is the amount of "hard news" in the publicity. For example, an April 1990 press release from Sarich Technologies Limited stated that the Orbital Combustion Process engine was already meeting the U.S. emissions standards not just for 1995 but those proposed for 2004 and beyond. Sometimes Sarich's penchant for making observations in front of the press on what he envisioned for the future got him in trouble because it was reported as already accomplished, but his two decades of public relations efforts have made practically everyone in Australia and everyone in the automotive world around the globe aware of the names *Ralph Sarich* and *Orbital*. And, for the most part, favorably.

Third, unlike many inventors, Ralph Sarich is a businessman. He understood running a company and making money from his pre-Orbital ventures. He allied himself at the outset with good partners in Tony Constantine, an engineer with operating business experience, and Henry Roy Young, a successful businessman and longtime friend. Enlisting the financial support and connective business tissue of Broken Hill Properties was an accomplishment. It was made possible not only by Sarich's conviction as to the importance of his invention but by his tough, businesslike approach to making a deal. The same qualities made him effective as head of the presentation team that took the Orbital developments to major auto manufacturers, where Sarich took the role of chief negotiator. He sold many sample engines at $1 million each to help keep the company moving before options and licenses began generating big money.

Fourth, Ralph Sarich has guts. Some might call it "nerve" or even "chutzpah." As mentioned earlier, when he concluded in his own mind that applying the Orbital fuel injection system to a conventional four-cycle motor was not revolutionary enough to force the major companies to use his engine, he turned down a major contract with General Motors. When Toyota and other manufacturers began to develop their own two-stroke engines, Sarich Technologies took pains to chastise reporters and educate the public lest loosely worded press reports lump lesser developments with the Orbital inventions. And, of course, establishing one's own motor manufacturing plant in the United States, even with some support from the state of Michigan, is the decision of a bold man, a man who absolutely believes in the uniqueness and protectability of his product.

Having described the remarkable achievements of Ralph Sarich in some detail, I must add a caution. There is a long history of automobile engine inventions from the spark plug to the intercooled turbo, and few of the "auto inventors" have remained heroes as time went on and the competition caught up. Remember Felix Wankel?

I have saved some stories of unusual invention-based start-ups for the next chapter, followed by the solid ground rules that all these successes suggest.

Chapter 13

Start-ups: The Common Ground Rules

Professional business skill—that's what separates start-ups that survive and succeed from those that move forward in fits and starts, or die young. One could do worse than to study the invention management and product launch skills of Dan Friel and others in these start-up chapters.

"Easy to say," you may think, "but is it a learnable skill or is it luck in a Brooks Brothers suit?" I believe it's the skill to learn fast enough to keep your business alive—so that you can then enjoy whatever luck comes your way.

"Take It in the Ear!"

Entrepreneurs who start up *two* invention-based winners in a row certainly deserve close study. Steven and Jon Lindseth come quickly to my mind. As recounted in Chapter 6, they made a huge winner of Interplak by combining their own business judgment with a team of top-drawer people with the engineering, manufacturing, and marketing experience needed to be at full speed from Day One.

As Interplak sped on its way, Steven Lindseth was involved in a venture group that included Ralph Lilore, a New Jersey lawyer specializing in technology licensing and financing, and Alan Mendelson, who ran Aetna's venture capital fund. To this group came a man named Jacob Fraden, a Jewish refugee from Siberia with world-class knowledge of the noninvasive measurement of body functions. Fraden had done stints at Case Western Reserve on biomedical work and then with Timex on medical instruments. Now he had discovered an infrared sensing mechanism, and the venture group set him up in a lab near Yale to develop products employing the technology. High on the list of applications was a thermometer that would take a human's temperature by placing an infrared sensor device in the ear for just a few seconds. No more glass thermometer under the tongue or in the anus. No more threat of spilled mercury.

A cinch to license, right? But real life worked its weird ways on their efforts. A deal was struck with Norelco, but before it could be completed the U.S. dollar declined so rapidly that Norelco, which imports all its products, had to back out. Teledyne Waterpik was seriously interested until a new president came in and said, "Hey, we don't belong in this market."

Jon Lindseth, who had invested along with son Steven, put it to the group: "Why don't I do the same thing we did with the toothbrush? Let's form a company and market it." Thus Thermoscan® was born. The Lindseths placed the company in San Diego because that's where most of the electronic thermometry companies were located, where experienced people could be found. They brought Jacob Fraden in to continue research and development and recruited top engineers from the IVAC division of Eli Lilly, engineers with knowledge specific to this field. They brought in William Krahel, a financial whiz, as president, and Peter Ellman, who had done a great job for Interplak, as VP Sales. John Hyle came to Thermoscan, Inc., from IVAC to build bridges to the medical community. And they recruited Betsy Winsett from Black & Decker as VP Marketing because, while the original Thermoscan would be a $299 unit intended for hospitals and doctors' offices, from the outset Jon Lindseth aimed for a consumer product to build the volume and profits needed to make the company a valuable future acquisition. Betsy would view Thermoscan as "package goods."

Once again, as with his Interplak company, Jon Lindseth required that the start-up executives buy stock in the company as evidence of their determination to succeed.

But before Thermoscan could launch a consumer product and go for big volume, the first goal had to be reached: acceptance by medical practitioners of a professional model offering a disposable probe cover. That required clinical evidence that a two-second reading in the ear was accurate, and that it was cost-effective. Independent clinicals and presentations at colloquiums led to publications in medical journals and to documentation of the model's accuracy. Fraden's patented pyrosensor device enabled Thermoscan to price the professional model competitively against electronic thermometers and far below the other infrared entry.

Hospital tests uncovered one serious problem. People liked these thermometers so much that they took them home and didn't bring them back. Thermoscan promptly developed an antitheft device. If the device isn't returned to its base for recharging every eight hours, a light starts flashing, "Return to Base!" and one hour later it becomes inoperative. This and other improvements raised the top-of-the-line price to $499, but the basic unit remained at $299. Sales in the second operating year reached about 14,000 units; sales of disposable probe covers—wherein lie the profits—began to climb geometrically.

The consumer model of Thermoscan, a simplified product but still a functioning two-seconds-in-the-ear temperature taker, was priced at about $100 in 1991 at the time of its introduction—with the firm intention of bringing it down in steps. Who would pay a hundred dollars for a thermometer that, after all, you use only once in a while? Time will tell. Meanwhile, I'm betting that there is a major market of families with feverish, bawling children who hate their parents for sticking a mercury thermometer in their dear little bottoms. And what proud new grandparent could resist Thermoscan, even at the cost of a C-bill?

Finding a Niche and Possessing It

When an inventor comes to see me, rhapsodizing over a wonderful new technology, as soon as I can get a word in, I ask, "Tell me, what product does this invention make possible?" If the inventor says, "Why, there are a thousand and one applications!" I shudder. Then I say, "Yes, but which one can be commercialized fast so that you can afford to explore the other thousand?"

Inventor/entrepreneurs, facing the many uncertainties of building a new business, do best when they remember this point and focus on the product, the distribution channel, and the allocation of money and manpower that will create a solid base and a defendable niche.

Niche marketing is frequently used to compete in consumer products. You target a specific use or a specific user group where your product has an inherent advantage. You avoid going head to head with the major competition. Sometimes a niche product—say, Grey Poupon Mustard—can be moved from a specialty for the cognoscenti to a mass product for the yupscale generation.

Niche marketing is especially useful for invention-based start-ups in which capital and organization are typically limited. Although the niche concept applies mostly to consumer goods, unique industrial inventions can succeed through this strategy too.

Here's how a boy from the Bronx grew up to do exactly that and staked out a place for his products in outer space.

Ernest Schaeffer went to Stuyvesant High School in New York, a school for bright kids with a scientific bent. For economic reasons, he never got to college and learned about things electrical as a TV repairman. Then, by applying his native talent to designing electric motors, he worked his way up through the engineering departments of Beckman Instruments, Skurka Engineering, and Whittaker Corp. At Whittaker, Schaeffer convinced his employer to take on a contract to develop a specialized motor for use in the space program, a project that developed into $750,000 in annual revenues, small potatoes by Whittaker standards but enough to give Schaeffer a chance to implement his unique ideas for specialized minimotors.

When a Whittaker corporate shuffle eliminated Schaeffer's special department, he licensed all his motor designs from the company and went into business for himself. He had no money or factory, only the good wishes of a handful of people at Whittaker and those in the small world of aerospace supply who were aware of his unique talent.

At first, it was hard scrabble, working out of one room at home. Next came a break, a job referral from his friends at Whittaker to design a solenoid for use in the tape recorders on Apollo II. Then Hughes tapped him to create an actuator, an extremely powerful miniature motor to control the telescope movement on the Pioneer 10 unmanned space probe.

Ernie Schaeffer knew he had found his niche. Such motors required not only clever design but absolute perfection in the selection of materials and construction because heat, cold, stress, and vibration are all the enemies of accuracy and durability. Anything short of 100 percent performance was unthinkable, not just to NASA but to Schaeffer. He ceased doing any other kind of work and concentrated Schaeffer Magnetics on 100 percent-reliable actuator motors, uniquely designed, each component conforming to his tight specifications. Soon Schaeffer became *the* source for such motors for manufacturers worldwide who make satellites or space machines of any sort. How could they take a chance by going elsewhere?

There is a bit of marketing advice from the Stone Age that was so well heeded by Ernie Schaeffer in the Space Age: "Find a need and fill it." What Ernie did and what you as an inventor/entrepreneur should do is to fill the need so well, stuff the niche so full of your special answer, that competitors can't compete and salesmen are unnecessary. Schaeffer Magnetics doesn't have even one. The product is the message!

Packaging Inventions

In Chapter 9, I pointed out how difficult it can be to commercialize a packaging innovation. Typically, the end user of the package tells the package inventor, "Yes, we might be interested if one of our suppliers were making it." And when the inventor talks to the package manufacturer, the response is, "If we had enough orders for this, we might go into the business of making it."

Some packaging inventions, however, lend themselves to a start-up scheme that unblocks this impasse. The key is to invent the machinery that makes the innovative package possible and then sell or lease the machines to either package makers or end users, depending on the economics of the situation.

A dean among the inventor/entrepreneurs who have exploited this start-up scheme is Sanford Redmond, whose dispenSRpak® machine makes a portion-size package for liquids and gels that you can open and dispense with one hand. The package has obvious advantages for single servings of mustard, ketchup, salad dressings, and other liquids that you normally find in a foil/film pouch—the kind of pouch that you open with your teeth and spill on your shirt in an airplane or a fast-food place.

Each dispenSRpak consists of two small plastic hulls—like twin side-by-side pontoons—with a flat plastic sealing sheet across the top. In the center of the top is a raised, elongated pyramid that has been scored along its elevated edge. To open the package, you just hold the package along the sides between your thumb and your third finger, putting the tip of your index finger between the two hulls, and pull back the two sides. The package opens at the scoring on the pyramid and the contents pour out in a reasonably controllable stream. In addition to its one-hand convenience, the dispenSRpak (as you may have guessed, the capitalized letters are you-know-who's initials) eliminates the frustrating problems of tearing open conventional single-serve packets, getting the contents more or less on the target, and occasionally getting the mustard or jelly on your tie or blouse.

Redmond was hardly a novice at the package inventing game when he started this venture. Not only did he have more than 500 issued patents in his name, he had also invented and in the early 1960s successfully commercialized the machinery for making those omnipresent butter pats—the little square pieces of cardboard carrying a slightly smaller slab of butter topped by a piece of wax paper. Most of America's butter patties, or "Reddies" as they are called in the dairy trade, are made on Redmond's machines, and the invention made him independently wealthy.

To the uninitiated, selling an innovation like dispenSRpaks would seem a lead pipe cinch. Who wouldn't prefer these neat packages that require only one hand and one motion versus the tough-to-tear pouches? What manufacturer wouldn't want to be first to put his product in this kind-to-consumers package?

Real-life marketing, of course, doesn't always rest on such rhetorical questions. The fundamental problem for dispenSRpaks was that consumers, who would applaud the packaging, had no influence over the decision to put products in them.

Institutional providers such as airline food suppliers are understandably married to a lowest-cost strategy. For products sold to the consumer at retail, such as foods or detergents or cosmetics, only major marketers can afford the equipment and the promotion costs—which automatically means that the sale must be made at many levels within the corporation.

Among the first takers was Mitsubishi in Japan, which resulted in dispenSRpaks being used for soft-drink concentrate. That product, ironically, was introduced at Disneyland in Tokyo. Unilever picked it up for its ketchup in Germany. In Australia, Master Foods Corporation is marketing a line of products under the name Squeezemates. Oscar Mayer began using the package for mustard in its Lunchables snack food packs.

As Sanford Redmond told me, within a short time after introduction, he had customers for fifteen of his dispenSRpack machines, and it seemed like an auspicious start. Priced at $350,000 per machine with an ongoing per-package royalty, he could visualize a repeat of his butter pat success.

However, these closings were only a small fraction of the companies that had expressed interest and an even smaller fraction of the potential market. What's more, none of the truly big-volume U.S. markets had been cracked. So Redmond began to focus on finding a customer with such high visibility that a successful sale would tell the whole marketing world, "Put it in dispenSRpaks!"

P&G was a clear target; its Tide® was an obvious product. A study of the detergent market told him that P&G was interested in selling a very concentrated Tide. Sanford Redmond put together a detailed analysis to show that sixteen dispenSRpaks in a sleeve would cost P&G less than its giant-size plastic container and would do as many loads of laundry, given a concentrated formula. Homework on P&G's organization told him that, while the sale had to be made at several levels, contact at the very top was essential. So he arranged a meeting with the president of P&G . . . and the results may be ready for the second edition of this book.

Redmond also conceived of his unique package as the catalyst for combining food products from more than one division of giant holding companies such as Philip Morris, and here, too, he is approaching people at the top corporate level.

At the other end of the customer range were the smaller companies that cannot afford a $350,000 investment. For these people, Redmond offered to lease the machinery to them with a royalty on each package produced. Redmond's comment on the relative merits of selling versus leasing is instructive: "I design the machine to suit myself, the inventor. I design the financial arrangements to suit the buyer."

I suspect he would *give* the machine to anyone who would guarantee him enough royalties from the volume of dispenSRpaks. And he's certainly in concord with the golden rule of invention marketing: Get it out there, any way you can.

Redmond has no forecast in the conventional sense of a business plan, but that doesn't mean he has no vision of the future. "Take it from me," he told me, "every major fast-food chain in America, every health and beauty company is testing them. These dispenSRpaks fit dozens of products. Butter pats were good for only one thing. This is the revolutionary food invention of the century."

Modesty aside, Sanford Redmond has the qualities that make for greatness in any business which, like packaging, must be sold to the final consumer as a part of another party's product.

These qualities include generating "instant impact" inventions. You see the dispenSRpak and you know it's clever, you know it gets rid of a commonly felt dissatisfaction. But equally, Redmond has the ability to implement his inventions—to do the homework on materials and cost, to refine the engineering so that everything runs right, and to demonstrate that the invention will create increased profits—or *de minimis,* not lose margins—for the company that converts to it.

The Sanford Redmonds of this world are, naturally, not commonplace. On the other hand, a number of other package innovators have used the "provide the machinery" approach. A good example of the latter is Specialty Equipment Corp., a Massachusetts company whose president, Fred J. Dowd, engineered a machine to make a folding corrugated box with bottom flaps that automatically lock into place when the box is opened so that they form a solid bottom without glue or taping. This little-known company sells its Automatic Loc Bottom machines for $500,000 to $600,000 each and has developed a world market.

Computer Start-ups: Hard and Soft

In the 1970s and even more so in the 1980s, new companies were started at an unprecedented rate in the United States to make and sell products that capitalized on the microchip and the information revolution. Twenty-seven out of the fifty start-up companies in this country that reached $100 million in sales in the 1980s were in computer-related businesses—hardware, software, networks, and retailing—and always a business that was swept upward by filling a need that, until its advent, either no one had perceived or could technically fulfill.

Take any list of the hot young growth companies for any year and you will find that about half of them are tied to the electronic information revolution. Fortunately for the founders of such companies, this computer-related zoom-boom created "overnight millionaires" because—with notable exceptions such as Apple, Microsoft, Sun, and some others—many such start-ups are hoisted on the petard of temporal technology.

For inventors or anyone investing in start-ups, the crunchy issue is tracking competition. Innovations in software and, to a certain degree, even in hardware do not require heavy capital equipment investments—which means that untold numbers of bright people around the country and around the world can be working on developments that will compete with or displace yours. Being there "fustest with the mostest" is always a good start-up strategy, and quite mandatory in a field with both a volatile marketplace and a new competitor hiding behind every computer.

There is a bit of philosophy I once saw on a wall plaque that seems to sum up the problem in the computer hardware/software business: "Just when I thought I had life all figured out, they went and changed the rules."

A Case of Not-So-Artificial Intelligence

Even though the computer world is gigantic and labyrinthine, millions are made. I want to describe one example of how level-headed basic business management can take a computer-based idea—an idea based on common sense rather than on artificial intelligence—and turn it into a potentially explosive profit maker.

The company is Pyxis, Inc.—pronounced, "Pix-us"—a name derived from the Greek word for a small jar in which medicines were kept. The company's principal product is called MedStation®, a computer-controlled chest of dispensing drawers that not only restricts access to each drug to a nurse entering his or her code and the appropriate patient's name but that also records the entire transaction, simultaneously posts the transaction to the hospital pharmacy's inventory control file, and automatically posts the correct amount to the patient's bill in the accounting office. It's sort of an automatic teller machine for medicaments.

As Ron Taylor, president and CEO of Pyxis, Inc., pointed out to me,

> The basic idea for our product originated with an M.D. who saw all the waste and billing slippage and narcotics-supervision problems in hospitals. His scheme was to control all of a hospital's supplies through a computerized dispensing system. But we did a series of interviews and focus groups with hospital people—and we found out that you have to change that world a little at a time to get acceptance. That's why we designed MedStation to serve just one nurses' station and sold our first hospitals anywhere from one to a dozen units.

Over and above doing so many things right from a marketing standpoint—like meeting a crying need for controls on drugs—Pyxis obtained patent protection by integrating common components with its own software to precisely fit hospital applications.

Taylor also recognized that he had to attack the financial implications vigorously. Naturally, he did studies to document the amount of nurses' time saved via MedStation versus hand-dispensing and record keeping. Also, Taylor and his marketing group were able to obtain hard numbers as to the upper and lower limits of the direct costs to hospitals of drugs stolen or not properly billed—numbers that any experienced hospital administrator would buy.

Equally important, Pyxis zeroed in on the principal difficulty in selling equipment to hospitals: capital expenditures. Then it devised a rental program for MedStation with a monthly price that was clearly lower than the existing losses from theft or slippage. That made the time saved for nurses and the avoidance of narcotics abuse simply bonus arguments. If there was no capital expenditure and a positive net effect on revenue from the first day of use, why not try it?

In a field that many consider extremely hard to penetrate, hospital equipment sales, Pyxis was able to sell more than fifty of the largest institutions in its first two years and to keep those customers by training and maintaining a service force that arrives on a hospital's doorstep within twenty-four hours to service or replace the machine. Formed by people with experience in the pharmaceutical field and in start-up financing, Pyxis became profitable in its third year.

I asked Ron Taylor how big he thought the market would be. "Over a hundred million dollars a year," he said, "and, yes, we have a plan to saturate the market before Burroughs or one of those guys can get in and, yes, we are just launching our second product for dispensing and recording supplies and, yes, we have a third one on the drawing board."

Pyxis is an example of keeping the focus on the most feasible opportunity and making that product irresistible to the customers who matter most. It's a formula that start-uppers should take to heart.

Is There a "Right" Way?

In the American educational system, and in the minds of many graduates of some of the best business schools, there is a shibboleth that all problems have answers. And the answers should be clear-cut, right or wrong. People have developed a multiple-choice mentality.

The art of invention does not work that way. As was the case with Sir Alexander Fleming's discovery of penicillin, success often comes only when inventors throw away their preconceptions and understand what they have *actually* uncovered, even though it is not the thing they were originally looking for. Similarly, many great inventions defy prior wisdom as to what will work or not work.

The business side of invention has this in common with the invention process itself: Success almost always requires going imaginatively away from the beaten path, the conventional answers.

David Montague found a way to finance his foreign bicycle production by getting letters of credit from his American distributors. Robert Dean created a combination licensing/franchising format for his concrete blocks. Ron Hickman found hard-scrabble ways to sell enough WorkMates to demonstrate that all the corporate executives who had turned him down suffered from marketing myopia. Ralph Sarich had the guts to walk away from a sure sale of a four-cycle engine when he decided that only a breakthrough in two-cycle engines would keep his company viable in the long run.

No matter how wonderful the invention may be, getting the company to start up and stay up requires a unique lift in financing or marketing or manufacturing, or all of them, not just writing a good business plan with beautiful spreadsheets but a unique execution.

Inventing Answers for Start-up Companies

Although invention-based start-ups share the problems common to most start-up businesses, certain issues rooted in the fact that you have not only an untested business but an unproven product need special attention. Here are the five questions to focus on:

1. *Who is going to run the company?*

• The inventor or originator of the product may not be the right person to run the business. But who is going to make that decision, and under what circumstances? And if the inventor or originator has to be superseded, how do you do it without breaking too many eggs? There is a case in point: When the board of directors of Colorocs felt compelled to remove Charles Muench as CEO, they set up a separate world for him in which he ran certain international operations; thus they kept him in the picture but not in the command chain.

• The person who runs the business at the outset needs to be strongly entrepreneurial in the sense of being able to do things that large companies cannot. For example, Jim Andersen, president of Applied Intelligent Systems, Inc., of Ann Arbor, Michigan, worked out a deal with Tom Nastas, president of Innovative Ventures, Inc., and managing partner of the Michigan Product Development Fund, to raise $700,000 to launch a new vision-system product by pledging a portion of the revenues from that product—in effect, a royalty. The rate of return for the investor was 25 percent with a five-year payout plus either continuing income thereafter or a $1 million buy-out. That's not just entrepreneurial, that's inventive.

• Gathering a quality team at the earliest possible moment is a hallmark of successful inventor/entrepreneurs. The most practical, and sometimes the only, way to attract strong people is by offering ownership—which may well be in the form of stock options exercisable at a time when productivity has been proved. Marc Newkirk of Lanxide has a fairly ideal policy. His people are paid salaries that are quite market-competitive, but instead of bonuses, which take needed cash, those who make contributions recognized by a broad-based management committee receive stock options. Although the company is not yet public, based on private sales of Lanxide stock, several employees of this start-up company are at least millionaires on paper.

• Phase II of the "who will run the company" question occurs when the business gets off the ground and reaches somewhere between $25 million and $100 million in annual sales. Frequently, the driver who broke the rules to jump-start the company isn't the person with sufficient management skills to run a larger organization. You may know this, but what are you prepared to do about it, especially if the person who started the company is you?

2. *What kind of company do you want to build?*

• Young companies, like young children, can be trained and dressed to fit the role you expect them to play. Do you want to drive this company into the Fortune 500, or do you want to build it fast and sell or merge it? Even if you change the objective later on, you can't waffle at the outset.

• All kinds of daily tactical decisions flow from that strategic objective. If you know that money must be made quickly before the competition catches up, don't invest in machinery and mortar. If a given piece of production equipment will make it difficult for competitors to copy you, it's not enough to buy or build that equipment. You must find a way to make it hard for others to get their hands on such a machine.

• At the outset, it's important to recognize the strengths and weaknesses of the team at hand. Shore up the weak areas by finding allies outside the company, people who will provide the power of a "big company team" even in the early stages. The fact is, investors, suppliers, and customers of invention-based start-ups bet on the people involved more than on any business plan or sales presentation.

• It's also important to get the business structure right from the very beginning. Even if you raise money through a limited partnership, you'll need a corporation as an operating vehicle. In which state you incorporate, how patents are valued as assets, how costs are written off, and other tax considerations—on all such matters professional planning should be in place from Day One.

3. *What is the real competition?*

• One of the most difficult things for the start-up entrepreneur is assessing where the competition is coming from and how much lead time is reasonably ensured. The fact is, it's hard to find out what may come up to bite you; but that doesn't diminish the importance of checking as seriously as possible. In some fields, like biotechnology, so much is happening in so many places that no one is completely sure where the leading edge actually is.

For illustration, in 1988 researchers at Nova Pharmaceutical Corporation were working on bradykinin-inhibitors to block the symptoms of asthma when they discovered—quite serendipitously—that certain of these drugs were extremely potent in the treatment of arthritis and other inflammations. And this discovery used a different mechanism of action from that used by the antiarthritics being researched by both Merck & Co. and Abbott Laboratories, Inc., as well as by a host of smaller companies around the world.

For two years, Nova did its preliminary research, determined the basis for its patent claims, and wrote a learned article for publication. Only in 1990 did competing companies learn of the Nova discovery.

• In watching out for competition, it also pays to look beyond the narrow confines of your own field. Consider, for example, the potential effect on makers of videocassettes (or on owners of video stores) of an invention that permits a vast choice of movies to be transmitted across phone lines, cable, or satellite. Patented by Explore Technology, Inc., this invention digitizes and compresses the electrical signal so that the amount of material transmitted per minute is increased almost 500 times, making a home service economically feasible. A customer could simply dial in for a movie and it would be zapped directly to his or her VCR in, say, three minutes, to be played back at a convenient time.

4. *How can you build credibility?*

• In the beginning, nobody knows you. Nobody knows your technology or product. Nobody has a strong reason to believe in you. Except, perhaps, the folks who are starting this venture with you, your accountant, your lawyer, and your mother.

Inventor/entrepreneurs, seized with the enthusiasm of the true believer, sometimes find it difficult to understand that the rest of the world does not understand or care about or have faith in their new-new product. That's why an invention-based start-up has no more important task than to build up its credibility.

• Test the devil out of product performance. Drag in an authority to bless the technology. Run a minimarket test or sell it door to door or by direct mail or through a catalog. Curry the favor of a newsperson in a position to break a major announcement story.

Whatever you do, don't just expect the business world to find your claims credible unless you create the gestalt that shouts, "Here is a winner."

No one has summed this up better than Alan Toffler, who manages a venture capital fund in California: "Financial people see hundreds of plans and they all say they'll make millions. Why should someone believe that what you have written is the truth? Answer that and you'll have a good shot at winning investors!"

5. *Can you take care of yourself?*

• People who start invention-based businesses, as you have learned, often wind up making millions of dollars. Others, unfortunately, end their adventure by losing their investment and/or their stockholdings—even though the company succeeds.

Bruce Verbauer put years into the development of the sponge female contraceptive that became VLI Corporation, which in turn became the darling of Wall Street for a period of time. But when marketing progress did not translate into bottom-line results, Verbauer was forced out. The inventor of Rollerblade, the in-line roller skates that boomed into an estimated $150-million-a-year industry, lost control of his company and he too found himself ousted from the CEO chair. He went on to develop a competing product, an interchangeable roller skate/ice skate, but found it tough sledding.

• The principle is clear, but it takes guts to stand by it. If you are an inventor/entrepreneur with the skills to operate a company, by all means do it. If, on the other hand, business management is not really your bag, hire top-drawer people to do it, but keep enough stock to ensure yourself a major voice in the business and a big payout when the company is sold or goes public.

For a good compact education on taking care of yourself, read *How to Run Your Business So You Can Leave It in Style* by John H. Brown with Irv Sternberg.

To Start or Not to Start, Always the Question

With all the advantages of exploiting an innovation that starting a company offers—greater financial potential and more control over your destiny—the alternative choice of licensing always deserves close review, both before and after a start-up.

Here are some start-up questions to ponder each time you think about this subject:

• Whose lunch are you munching? Will you awaken giant competition before you are entrenched?

• How much staying power do you really have? Is a second round of financing committed and in place before you launch the first product?

• What kind of life cycle do you foresee for your product? Does it have instant impact? Does it have news value, or does it require slow education? And what will you do for an encore?

• Is the product truly market-ready? Or does it need the kind of production engineering, packaging, or distribution that only a large company can deliver? What are the chances that O'Brien's Law will operate before you get to market?

• Is there a company out there that needs your innovation so badly that it will make a license deal ensuring you income equivalency with a start-up but without the risks?

• In addition to starting this company, can you also license one or more corporations in the United States or abroad to help finance your operation? And if you're not ready to do this at the onset, then when?

There are more questions you might ask, but these make the point: should you license, start up your own company, or both?

Licensing, relatively, is the "easy" way to money.

Starting your own business is one of the most wonderful, exasperating, worthwhile, challenging, frustrating, growing, enriching, or impoverishing experiences you can have in a free society. But if you have an invention with a raison d'être, there may be no other way to realize real rewards.

The innovator may envision a future Fortune 500 company, while the rest of the world sees only a mass of unknowns. The task of the inventor/entrepreneur is, above all, to convert his own missionary enthusiasm into a coterie of enthusiasts. Only those with fierce determination and fiery imagination need apply.

Chapter 14

"Together or Separately"

It was, of course, Ben Franklin, a man of many talents including a penchant for invention, who said at the time of the signing of the Declaration of Independence, "We must all hang together, or assuredly we shall all hang separately."

If the United States is to retain its position as a preeminent industrial power, there is an urgent need to apply this principle to the creation of new technology and new industry.

Much of this book has been devoted to explaining how those who invent can overcome the inertia and risk avoidance of those who have control of capital and corporate structures. In the last few decades, many people have made millions from inventions that not only improved the quality of life but also created new sources of both profit and employment.

In the same time period, however, many millions more were made by ingenious minds that focused on corporate takeovers and financial restructuring, which, by and large, redistributed wealth but neither created new jobs nor brought in new internationally competitive technology.

The time has come for a merger. The minds that can see the inventions of the future must "hang together" with the minds that can manage the business side of invention.

Having devoted the greater portion of this book to viewing invention from the standpoint of the individual, let us now turn the spotlight on corporate America and on what corporate management can do to "hang together" with inventors both inside and outside the corporate family.

The Human Side of Invention

Wouldn't it be wonderful if we could unfailingly identify winning inventions and discoveries at the earliest stages, back the beauties with maximum money and manpower, and kill the fatally flawed innovations before they chewed up big chunks of our resources?

Well, we can't quite do that. But in talking with inventors who have converted their creativity into cash and to entrepreneurs who have turned the inventions of others into going businesses and to corporate managers who have scored consistently with new products, some useful indicators emerge. Call them the WINvention Ground Rules:

1. *The best inventors are pig-ass stubborn about what they are doing and, especially, about the feasibility of their discoveries.* What would be insufferable obtuseness in the typical businessperson is simply a creative mind seeing a solution so clearly that it will brook no opposition. Sometimes this is masked by diffidence and politeness; sometimes it is rank brusqueness. Sometimes, of course, "dead certain" turns out to be "dead wrong." No matter. Without this quality of fierce belief in the functional value of the invention, the chances of creating great monetary values are slim.

2. *The best inventors—and the business teams around them—need to understand what motivates the creative mind and to act on that knowledge.* There are only three motivators: the challenge of the work itself, the honor and satisfaction of succeeding, and money.

Perhaps because of their academic training, many engineers and chemists choose to think of themselves as "scientists" or "research & development" people rather than as "inventors." It may be that the image of Rube Goldberg, the cartoonist who created world-class complicated devices to solve everyday problems, hangs over the word *inventor.* Yet of all the successful inventors I have met, I have never known one who did not, after the fact, think that his or her development should be publicly recognized, perhaps even entered in The National Inventors Hall of Fame.

It may be that a history of disappointment affects inventors' financial expectations. In Tom McMahon's book *Loving Little Egypt* (Viking Penguin, 1987), Mabel Bell, Alexander Graham Bell's widow, sends Mourly Vold, the natural genius who has broken into and improved reception on long-distance networks, to study with electrical scientist Nicolai Tesla. Tesla reads the letter of introduction and tells Vold, a.k.a. Little Egypt, "You're welcome to stay with me as long as you like, and learn whatever I can teach you. I should tell you at the beginning, however, that there isn't any money in being an inventor. The boys at the bank get it all in the end."

Nonetheless, as this book demonstrates, there *is* gold in invention, and money can be a powerful force in moving inventors, although love of the inventing process is indispensable too. And inventors who aim for wealth should not be surprised that those who can create rich profits from the invention will share the proceeds.

3. *The best inventions, the really explosive ones, come from one of three directions:*

• A crying need, perceived and attacked until the invention emerges, such as the use of AZT for the treatment of AIDS;

• A discovery, whether serendipitous or deliberate, applied to a marketing opportunity, such as Wilson Greatbatch's realization that his electrical stimulator invention would be a perfect heart pacemaker; or

• An application of what is already known to a new purpose, such as the use of Goodyear's vulcanized rubber not just for tires but for replacing hobnailed heels on shoes.

Wisdom for Independent Inventors

In the course of talking with the inventors presented in this book, no one impressed me more with his professionalism than did George Clemens, inventor of the Interplak toothbrush and many other products. He has always felt that he could invent *anything*. But he didn't want to waste his time on products for which there was no genuine need. So he would go around the house, looking first in the kitchen to see how well everything worked, then in the bathroom, then around the rest of the house. Next he would look at consumer ads for products that might not live up to their claims.

One result was an understanding that toothbrushes didn't clean very well at the gum line. Clemens then personally checked all the prior art in the Patent Office, especially electric brushes—and there found the focus for his invention. The bristles of earlier electric toothbrushes wouldn't stay in the pocket between the teeth or push down in the pockets at the gum line because their unidirectional rotation spun the bristles out. The solution was clear: constantly reversing, counterrotating bristles. Then all that remained was to create a practical embodiment of the idea, patent it, and license it. After rejections from forty-two companies, George Clemens found the Lindseths, father and son, who moved Interplak to market and eventually put millions in the inventor's bank account, specifically about $2 million a year.

Clemens's story is a précis of what it takes for the independent inventor to create a worthy invention and to convert it to cash.

Thinking Critically About Inventions

To understate it, making any money, let alone millions, from invention requires a lot of thought. Traps are everywhere. The only way to avoid them is to ask yourself, "What is the worst nightmare I could have about this invention?"—and then figure out what you would do if it came true. What if the product costs 10 percent more to manufacture than you thought it would? What if the perfectly written patent application is denied—or reexamined? What if no company will license the invention even though it is absolutely logical that the companies you have approached will benefit from it and can readily afford it?

Not asking that last question, or, more accurately, not answering it properly, cost me some time and money on a nice tennis racket invention. I started playing tennis with a wooden racket. In those days, if you wanted to see if the net was the right height at the center point, you stood the racket vertically at the center of the net, put your finger at the top of the racket, and then placed the head of the racket sideways above your finger. The length of the racket plus the width of the head were exactly 36 inches—and you adjusted the net accordingly.

When the Prince® and other oversize rackets hit the market, the old net measuring trick went down the tube. I invented an answer: the "Net-Mark," a line on the edge of the racket frame placed exactly so that the length of the racket plus that portion of the width of the frame below the mark combined for the 36 inches. Now this invention would cost a racket manufacturer nothing, not one penny of added production costs, because it could be silk-screened on the racket simultaneously with the normal decorative pattern.

I thought every one of the racket makers would love this. I went to Prince, Wilson, Dunlop, and others. I was wrong. Nobody believed that this little advantage would determine a brand buying decision. Quite possibly, they were right.

The story had a P.S. I had filed a patent application and just as it was coming to judgment, a patent covering essentially the same invention was issued to the late An Wang, founder of Wang Computers and holder of forty-some patents and twenty-three honorable degrees. In the light of the universal yawn from potential licensees, it hardly seemed worth my while to file an interference and argue the date of invention.

Wisdom for Corporations

Imagine the CEO of a manufacturing company standing there at the year-end board meeting and saying, "Members of the Board, I am happy to report that we have doubled the number of new products brought to market last year compared to the year before—and we have even more new product breakthroughs ready for this coming year."

Unfortunately, such words are not being heard in many boardrooms. As corporations have grown bigger, as science has grown more complicated, as government regulation has grown more byzantine, as markets have splintered, as selling power has shifted from manufacturers to distributors, finding technology-based new products—products that can create a market and hold it—has become ever more elusive for American corporate management.

In most large companies, during the ever-upward 1980s, the policy was to throw more money at R&D . . . and pray for early results. New laboratories were built, new echelons of R&D people were layered, new controls installed, and more and more reports were written. Strategic objectives were set so that the white-coated people's workdays would be channeled to the corporate targets and little else. Only in pure research, a luxury for most companies, did the individual human mind enjoy the freedom of uncontrolled exploration.

But a funny thing happened on the way to invention. For most companies, the cost per marketable new-new product went up while the number of such products went down. And inventors from outside the hallowed walls often found that N.I.H. (Not Invented Here) had escalated to "N.I.M.L., Not in My Lifetime!"

Have I overstated and oversimplified the situation? Perhaps. But only to focus your full attention on the problem and the opportunity that exists.

Company managers in almost every industry face the problem: unprecedented competition, not only from Japan and the Pacific Rim countries but from a powerful European Community and from emerging Eastern European countries as well. It is a time that calls for U.S. companies to maximize innovation. To do business internationally, indeed to make the best strategic alliances internationally, superior technology and product performance are the critical cards.

Fortunately, invention is alive and well in America; you've seen that from the stories in this book, and I have, in fact, a file full of other cases. By opening the door to the power of the individual mind, corporate managers, seeking to step up sales and profits without straight-line increases in R&D, can capitalize on the same underlying conditions that made possible the birth of Apple Computer, the Kearns windshield wiper, or the Interplak toothbrush. The objective is to vastly increase the flow of new product opportunities both from within the company and from outside. Increased choice as to where to put capital investments improves the odds on choosing a winner; it's the same principle that enables large racehorse stables to pick colts that win the Kentucky Derby, the Preakness, and the Belmont and sometimes the Triple Crown.

Reaching In

The best way for a company to improve the breed of its new-new product opportunities is to create and implement a companywide policy that says, "Invention is a top priority for us." Or, as one company president put it to me piquantly, "If you're not the lead dog, the view never changes."

Here's what some companies have done to avoid having to play innovation catch-up:

• The 3M Company encourages its people to take up to 15 percent of their time to pursue innovations of their own choosing. An annual awards banquet is the occasion for 3M people to be publicly recognized for such accomplishments. Is 3M the best of all possible worlds? Of course not. It still has the problem of sheer size and of individuals with personal priorities. When I asked Art Fry of Post-it fame how he was so successful in moving inventions through the corridors of power, he answered ingenuously, "It's not too hard. You look for friendly faces. And you steer clear of people who can stop you."

• The Du Pont Company, cradle of major chemical inventions, has established several innovative channels for bringing fledgling ideas to fruition. In 1986, after an "innovation audit," Du Pont decided that many of its employees, not just the R&D people, had potentially valuable ideas. Du Pont set up a program called $EED, which offered grants of up to $50,000 to any employee, regardless of level or function, whose idea and commitment passed the $EED committee's scrutiny. Significantly, pursuing the idea did not cost the employee his or her job because it could be pursued part-time or on a leave of absence. Furthermore, assistance from knowledgeable Du Pont specialists was provided without cost. In a little more than three years, the $EED program received some 400 applications and made grants of about $2 million to 170 applicants. Dr. Richard H. Tait, who runs the program, says it is too early to make a cost-reward analysis, but the potential is clearly great and the effect on employee thinking is very positive.

Du Pont also runs a "School of Intrapreneuring" and provides senior executives' advice to those who would begin a new venture within the company.

• A number of other companies have tried "intrapreneurship." Few report great success, but this may result from underfunding or impatience. The reasons may go even deeper. Harold S. Geneen, once chairman of the gigantic ITT Corporation, maintained in his book *Managing** that intrapreneurship cannot work in large corporations because (1) the risks of entrepreneurial deals go against top management's need to ensure stable growth in order to satisfy stockholders—which is to say, Wall Street—and (2) true entrepreneurs stay with a company only long enough to gain experience and then leave to get the cash. Companies like Eastman Kodak, which ran through more than a dozen intrapreneurial businesses during the 1980s without success, have learned that being half-free is not a workable system for genuine entrepreneurs.

Other companies, among them Lubrizol, SmithKline Beecham, and Xerox, have ventured into various kinds of sponsorship for embryos. Results suggest that a structure that lets inventors create and lets entrepreneurs make their own decisions works best. Smaller companies have done it, too. Minneapolis-based McLaughlin Gormley King, Inc. (MGK) set up Jim Hardwick, an ingenious creator of specialized chemicals, in a separate corporation with a big share of the action. MGK kept strategic liaison but let Hardwick operate. In five years, MGK sold the company and made $15 million on the deal, ten times its investment.

• There are two ways that corporations can best capitalize on the individual's drive to "do it my way." One is to set such people up with their own operation—a floating subsidiary or joint venture. The other is to allow them a large measure of freedom to invent, even within the structured organization.

At General Electric, starting back in the 1950s, finding a method for producing synthetic diamonds became a high-priority project. The company formed a team of researchers, a group of dedicated scientists who later became known as The Diamond Fever Gang. The original team included H. P. Bovenkerk, F. P. Bundy, R. Cheney, H. T. Hall, A.

*Written by Geneen and Alvin Moscow, published by Doubleday & Company, 1984.

J. Nerad, H. M. Strong, and R. H. Wentorf, Jr., who were later joined by R. C. DeVries and T. Anthony. (Note the big-company style of initials instead of first names!) They were, I was told by Irv Begelsdorf, a crackerjack science reporter for the *Los Angeles Times,* rugged individuals, as diverse as their names suggested. What they shared was the grip of Diamond Fever, the pursuit of a commercially practical method of producing synthetic diamonds, not so much for the gem diamond market but for the even larger industrial market.

After talking with several members of that team, I deduced that General Electric's management had pulled off a great coup. To achieve the economical production of synthetic diamonds, there were two obstacles: identifying the right temperature, pressure, and chemical catalyst, and building a machine that would consistently replicate the conditions. So GE assembled these people with diverse skills in materials science and diverse personalities, not really a balanced team but a group of stars, all fascinated by the prospect of beating Mother Nature's most glittering achievement. And GE paid them with a currency that, for some scientists, equals money: freedom. After meetings and discussions of the problems of diamond manufacture, each scientist or pair of scientists went off in a unique direction. By early 1955, Bundy had achieved an "anvil" chamber that worked and Hall and Strong had developed a "belt" method that worked better. And by 1957, Bovenkerk had set up the first production of diamond abrasive grains at GE's Carboloy plant in Detroit. Everyone in the group had worked day and night to produce the first twenty-five carats that convinced management to proceed.

Significantly, one man tended to be a loner; Wentorf liked to do his experiments off in a corner. His range of investigation swung wider than the rest. Among his achievements was making a 1.3-carat crystal of gem quality, a patented method that taught the world to make diamonds that even Carol Channing would envy. Sumitomo of Japan now routinely produces crystals larger than 11 carats and DeBeers has used Wentorf's process to grow a 14.2-carat yellow crystal. Of even greater importance to GE, Wentorf also discovered how to convert the graphite form of boron nitrate into cubic BN, the second hardest material in the world, a remarkable grinding substance trademarked Borazon by GE.

When I asked Robert Wentorf what General Electric had done to reward The Diamond Fever Gang, he said, "They gave us salary increases and they gave us sort of little leaves of absence, and they'd give us a little money for patents, but that was only a token, you know, like you'd get a hundred dollars or so. They gave you more freedom... I don't know, they'd let you do what you wanted to do."

I've gone into detail about The Diamond Fever Gang because the story illuminates an important aspect of what motivates many inventors. The best people want to do what they want to do. They'll sacrifice income and perks for freedom. Please note, however, that this is not put forth as evidence that financial incentives are not equally or more effective. Not to mention the combination of both motivators. Computer companies and biotech companies commonly succeed by balancing monetary and intellectual incentives. Management at many large corporations, afraid of an adverse reaction from other employees, shies away from major incentive packages for inventors—even though it may let a star salesman make more money than the company president based on a commission arrangement.

Hyperstimulation of invention along with a better structure for intrapreneurialism are areas that cry out for more creative thinking from American corporate management.

Taking the Fear Out of New Products

One of the most important steps that management can take is to allow people to take responsibility, to give them the authority both to succeed, with defined rewards, and to fail without fear of hanging—at least, not the first time around.

At The Wellcome Foundation in England, Trevor Flanagan, one of its most astute licensing officers, summed up typical new product procedure this way: "In most companies, new products are the work of committees, and people like it that way... because when only one individual becomes the moving force, if the product fails, it's his head."

Another interpretation of the same syndrome was this drollery that I found on a corporate bulletin board:

The Six Stages of New Product Development

Stage 1: Euphoria
Stage 2: Disenchantment
Stage 3: Disassociation
Stage 4: Search for the Guilty
Stage 5: Punishment of the Innocent
Stage 6: Promotion of the Uninvolved

The concerns that underlie such black humor are not funny from the viewpoint of a farsighted management. People who 100 percent risk avoiders are already dead. The quandary is how to encourage the capable and ambitious and to avoid—or at least control—the incompetents. Esther Dyson, in her column in *Forbes,* neatly summed up the dilemma of the entrepreneur-turned-CEO:

> To get an idea off the ground, you have to be fanatically committed to it and persuasive enough to get investors and employees to go along. For an entrepreneur, the way to become successful is to build an institution around an idea, but the way to stay successful is to come up with new ideas—which your institution will probably reject.

From companies that succeed in nurturing invention I have distilled a close-to-ideal structure that I call the leading-edge group. This is how it works:

First, you create a separate unit of R&D people who are asked to invent new-new products, not variations on existing lines. Give them boundaries and goals, but be willing to look at any genuine discovery, even if it may be sold off or licensed out later on. Separate these folks physically from those who are creating improvements on current products or line extensions. Then hook up the inventor-type R&D people with marketing people who interface with them to give the innovations market relevance and who can actually take the output to commercial tests. Add one or more people from the manufacturing department, either as special liaisons or as employees on detached service. Then appoint a "president, new businesses" to run the combined group.

Make it an honor to belong to this leading-edge unit. Establish significant financial rewards that are tied not only to getting new-new products to market but to profit contribution. And provide an honorable exit back to everyday R&D or existing brand marketing for those who feel they aren't making it in the leading-edge group.

There are qualifiers. The minimum time frame for measurement should be three to five years, the minimum return on investment for incentive payments, 20 percent more than the existing company average. And there should be a provision for the company to exit by selling to its employees (or to third parties) any proprietary rights that have been rejected for development. And when a big winner comes along, give the people who invented it the opportunity of going to market with it.

Many variations are possible to suit individual situations. The key element, the one that makes all the difference, however, is the separation of the leading-edge group into an independently run unit. There's no lack of aggressive executives to run such operations. When you look at the number of entrepreneurial company executives who have left their organizations to head start-ups—for example, Rod Canion who left Texas Instruments to found Compaq Computer Corporation—you have to believe that there are leaders in many corporations who can launch new businesses from a leading-edge platform.

Stretching the Outreach

Any company president who believes that his R&D department has a corner on all the good ideas in his field is—well, in need of quick treatment for self-delusion.

As a matter of fact, very few CEOs believe this. What they often do believe is what their lawyers or R&D chiefs may tell them about reaching out for innovations. The lawyers warn of lawsuits from mad and misguided inventors. The R&D chiefs cry that they almost never see anything useful from outside and, in any case, their time is fully booked so they won't be able to deal with something new this year.

Is this fiction or fact? Well, it's both. Some companies absolutely cut themselves off from outside ideas. Here's a letter to an inventor from a very large toiletries company:

> Thank you for your recent letter regarding your proposed product concept. While we are always very interested in new ways to meet our customers' needs, it is our policy not to accept any new ideas that come to us from sources other than our own.

At the other end of the spectrum are companies that have departments to "license in" inventions. In an effort to avoid N.I.H., some companies even use independent technical evaluators rather than their own R&D people, and, as mentioned earlier, Harry Hoyt, CEO of Carter-Wallace, went so far as to pay his VP R&D a bonus for acquiring technology from outside inventors.

For most companies, however, reaching out, really stretching out to find innovations, is simply not done. But there are exceptions that light the terrain.

In the pharmaceutical industry, where licensing and cross-licensing is a way of life, many licensing officers do establish channels of communication with their counterparts around the world and with licensing officers and researchers at universities.

Even more surprising and worthy of study are the outreach programs of GKN Automotive, Inc., of Auburn Hills, Michigan, a maker of automobile parts, and Masco Corporation of Taylor, Michigan, which owns some 100 companies, including Delta Faucets, Baldwin Brass, a number of oil-tool companies, and a large interest in Henredon furniture.

GKN Automotive, a division of GKN Technologies, Ltd., a British company with almost $3 billion in sales worldwide, ran a program called Project Extra. The scheme originated in England, where the ratio of submissions worth investigating was one in a hundred and where the parent company licensed five valuable innovations in the first four years of outreach.

How the program was jump-started is interesting. Under the direction of the full-time program administrator, Sally Korth, a series of formal presentations was made to inventor groups, the venture capital community, and state and local government groups that work with entrepreneurs. A brochure and a press release followed, explaining GKN's fields of interest and making it clear that submissions of patented or patent-applied-for innovations were preferred.

In the United States the program was launched in 1990 and in the first year generated about 150 submissions, of which several had been selected for evaluation. Unfortunately, bad times in the auto industry led to budget cuts that have temporarily suspended Project Extra.

At Masco Corporation, Dennis O'Connor, director of new products and technologies, runs the outreach program in the role of an activist licensing officer. He has established a network of contacts to whom he has let it be known that Masco is receptive to outside invention. He does this by attending meetings and carrying on an extensive correspondence. His targets are patent attorneys, the Inventor's Council of Michigan, various state and local organizations that offer support services to inventors and entrepreneurs, and the venture capital community. Because of the wide diversity of products made by Masco's various subsidiaries (whose sales total $3 billion), O'Connor's outreach attitude is, "If it's a great invention, bring it to me and I'll figure out where it fits in our group of companies."

O'Connor also volunteered some helpful hints for making an outreach program succeed:

> "I have people to work with in every company who are not purely R&D. When I submit something to them, I know that, instead of feeling threatened, they'll give me a nice, honest evaluation. Also, we're very flexible on terms, with lots of different arrangements from royalties to buy-outs. If

we really like something, we'll even undertake the patent filings ourselves."

In sum, despite the obvious risk of claims or lawsuits from inventors with either real or imagined complaints, it would seem to be good business for more company managers to turn their best legal and public relations brains loose on devising a system for looking into the inventive minds beyond the corporate gates.

As a starting point, one might study the toy industry. There, even the top companies depend upon outside inventions in lieu of a bigger internal product generation. Mike Lyden of Tyco Industries told me this:

> The receptivity of toy companies to outside innovation has led to a coterie of 150 to 200 professional inventors who work full-time on toys and games. We have twenty-five to thirty of them that we deal with regularly. We don't have the time or the people to handle the paperwork for every submission that comes out of the blue, although we may make an exception when we have a specific reason. But we need only five or six new products a year and we look at over a thousand.

Because there are over 100 companies in the toy business, obviously there are enough professionally generated inventions to satisfy the industry and to provide a marketplace for inventors.

Six Steps to Dialogueland

For company managers looking to open useful dialogue with the owners of new products and technologies—profit opportunities created with other people's money—here are six minimum-risk moves you can make:

1. *Start by spreading the word, first to your own employees, then to the world around you, that your company is receptive to outside inventors, that the welcome mat is at the front door.* Whisper it or shout it; just do it. Some years ago, International Playtex ran a good-size advertisement in *The Wall Street Journal* with the headline, "Multi-million Dollar Company Seeks New Products." It received a flood of replies, some of which were wending their way through development when the company was visited by the Angel of Reorgani-

zation. You need not go to national advertising to start. Low-key PR, mailings to appropriate groups, even asking your own R&D people to recommend their inventive friends—there are many ways to start a flow of ideas large enough to generate a few nuggets.

2. *Adopt a friendly corporate voice.* Sound friendly both in person and in documents. What's the use of a legal release form that oozes hostility and says in effect, "If this idea were any good, we would already have thought of it?" Far better to precede an initial meeting with a simple letter. Here's the essence of one I used for a meeting with the Clairol Division of Bristol-Myers Company:

> This letter will confirm our agreement that during the meeting to be held on [date], no confidential or proprietary information will be disclosed. In view of the nonconfidential nature of the meeting, neither party shall incur any liability to the other for having received any information during the meeting, whether such information be oral or written.

In addition, communicate to your people who meet with outside invention owners that friendliness is cost-free. When it comes to asking hard questions, ask them in a soft voice.

3. *Guard your reputation for fairness.* Companies that consistently "go to school" on inventors without buying or licensing anything do themselves a disservice. In every field, there is a coterie of experienced inventors, and if these people get "down" on your company based upon repeated unpleasant experiences, no one benefits. "Adult education," learning from the other guy, is a normal part of business, but using ideas disclosed in confidence is poor business. Few companies encourage this, but because your company can be hurt by the actions of a few individuals, it is wise to protect your company's reputation by having a clear written policy on this point.

4. *Put a broad-gauged person in charge of screening invention submissions, one who not only appreciates technology but also has strong marketing insights.* After all, it's the explosive combination of something that works better with a powerful, preexisting need for that item that pays off big.

Sometimes, a big technology dictates the market. Other times, it's a creative application of the invention that makes it worthwhile. An extreme example is the whole panoply of software created each time chip capacity increases.

Theoretically, the inventor should know the best use for his discovery. But experience shows that this often isn't the case. The businessperson who can spotlight the best marketing fit will change the value of the discovery. Make sure your screening person has that skill.

5. *Be tough in looking at outside innovations.* Look for the opportunity first; but promptly, after you decide you're interested, look for the fatal flaw. After all, this innovation was developed with someone else's money. So far, so good. But the next money, development money and marketing investment, will be your cash. So try, early on, to find the unfixable—not the nits, just the unfixable.

You can't afford to be tough unless you look at many opportunities. In most businesses, an outreach program must bring the company fifty or more qualified innovations a year before you can be tough and still emerge with one or two good opportunities.

6. *Form a partnership with the inventor.* Give him or her a generous option payment during the period when you're evaluating or preparing for market. In return, make the inventor responsible for either work or consultation to ensure that matters don't get derailed. Above all, maintain management-level communication with the inventor so that if he or she believes that R&D or marketing is going off the track, that opinion can be reviewed. Nobody is right all the time, even a brainchild's parent; but it pays to listen to someone whose heart and pocketbook are in the same place.

For independent inventors and for the creative minds in the corporate world, I have a happy thought that you can take away with you: The opportunity for profitable invention is alive and well. This is not just the idle talk of a Pollyanna but something you can observe for yourself by looking about the world and seeing what is desperately imperfect—and inventing the solution.

As evidence of this, I call your attention to current restraining devices for automobiles. How do you like seat belts? Personally, I detest them. They're uncomfortable. Sometimes they interfere with driving. And the kind that are attached to the door of the car can choke you to death. "Oh, yes!" you say, "but what about air bags?" That was my reaction, too, until I found out that the Department of Transportation and the auto manufacturers all say that you should wear your seat belt, even if you've bought air bags!

Well, there it is. Go to work.

An Inventor's Bookshop

Turning an invention into a fortune requires many kinds of knowledge, as this book demonstrates. While there's no substitute for professional specialists, here are the most useful books I've found for anyone seeking information about the business side of invention.

Roots of Invention

The History of Invention
Trevor I. Williams
Facts on File Publications, New York
1987

The Smithsonian Book of Invention
Smithsonian Exposition Books
W. W. Norton & Company, New York
1978

The Way Things Work
David Macaulay
Houghton Mifflin Company, Boston
1988

Information on Patenting

General Information Concerning Patents
U. S. Department of Commerce
Patent and Trademark Office
Superintendent of Documents
Washington, D.C. 20402

Inventing and Patenting Sourcebook
Richard C. Levy
Gale Research, Inc., Detroit
1990

Inventor's Guide to Successful Patent Applications
Thomas E. DeForest
Tab Books, Blue Ridge Summit, Pa.
1988

Patent It Yourself!
David Pressman
Nolo Press, Berkeley, Calif.
1990

Patent Pending
Richard L. Gausewitz
Devon-Adair, Old Greenwich, Conn.
1983

Protecting Intellectual Property

Copyrights, Patent & Trademarks Worldwide
Hoyt L. Barber
Liberty Hall Press, Blue Ridge Summit, Pa.
1990

The Executive's Guide to Protecting Proprietary Business Information and Trade Secrets
James H. A. Pooley
Probus Publishing, Chicago
1987

How to Protect Your Creative Work
David A. Weinstein
John Wiley & Sons, New York
1987

Intellectual Property
Arthur R. Miller and Michael H. Davis
West Publishing, St. Paul, Minn.
1983

Patents, Copyrights & Trademarks
Frank H. Foster and Robert L. Shook
John Wiley & Sons, New York
1989

Who Owns What Is in Your Head?
Stanley H. Lieberstein
Hawthorn Books, New York
1979

Commercializing Invention

How to Turn Your Idea into a Million Dollars
Don Kracke with Roger Honkanen
New American Library, New York
1979

A Handbook for Inventors
Calvin D. MacCracken
Charles Scribner's Sons, New York
1983

Idea to Marketplace: An Inventor's Guide
Thomas R. Lampe
Price Stern Sloan, Los Angeles, Calif.
1988

Invent and Get Rich
Ernest A. Zadig
Prentice-Hall, Englewood Cliffs, N.J.
1985

Inventing—Creating and Selling Your Ideas
Philip B. Knapp
Tab Books, Blue Ridge Summit, Pa.
1989

Inventing for Fun and Profit
Jacob Rabinow
San Francisco Press, San Francisco, Calif.
1990

Inventions—How to Be Successful
George Molyneux
International Thompson Business Publishing, London, England
1989

Inventor's Marketing Handbook
Reece A. Franklin
AAJA Publishing, Chino Hills, Calif.
1989

That's a Great Idea!
Tony Husch and Linda Foust
Ten Speed Press, Berkeley, Calif.
1987

Negotiating

The Art of Negotiating
Gerard I. Nierenberg
Pocket Books, Simon & Schuster, New York
1989

Getting To Yes—Negotiating Agreement Without Giving In
Roger Fisher and William Ury
Penguin Books, New York
1983

Entrepreneurship

The Encyclopedia of Small Business Resources
David E. Gumpert and Jeffry A. Timmons
Harper & Row, New York
1982

Entrepreneurship and Venture Management
Clifford M. Baumback and Joseph R. Mancuso
Prentice-Hall, Englewood Cliffs, N.J.
1987

Franchise Bible
Edward J. Kemp
Oak Press
1990

How to Run Your Business
So You Can Leave It in Style
John H. Brown with Irv Sternberg
AMACOM Books, New York
1990

One Step Ahead—The Legal Aspects of Business Growth
Andrew J. Sherman
AMACOM Books, New York
1989

Pratt's Guide to Venture Capital Sources
Stanley E. Pratt
Oryx Press, Phoenix, Ariz.
1988

Venture Capital Handbook
David J. Gladstone
Reston Publishing Co., Reston, Va.
1988

Business Plans

The Arthur Young Business Plan Guide
Eric Siegel, Loren Schultz, Brian Ford, and David Carney
John Wiley & Sons, New York
1987

Business Plans That Win $$$
Stanley R. Rich and David E. Gumpert
Harper & Row, New York
1985

New Product/Technology Management

Innovation: The Attacker's Advantage
Richard N. Foster
Summit Books, New York
1986

The New Products Handbook
Larry Wizenberg
Dow Jones-Irwin, Homewood, Ill.
1986

Pioneering New Products
Edwin E. Bobrow and Dennis W. Shafer
Dow Jones-Irwin, Homewood, Ill.
1987

Technology Management: Law/Tactics/Forms
Robert Goldscheider
Clark Boardman, New York
1988

Winning at New Products
Robert G. Cooper
Addision-Wesley Publishing, Reading, Mass.
1986

The Inventors and Their Inventions

Important Note: Some inventors, corporations and business people appear in more than one place in this book because their stories are good—or bad—examples of several different topics. Italicized chapter numbers indicate principal stories.

Name/Brand	*Invention/Subject*	*Chapter(s)*
Bennett, Robert	Plastic products, packaging components	2, 12
Beresin, Jack	Business advice	*4*
Berger, Dick	PLAX	*1*, 5, 6, 10
	Carpet Fresh	6
Berman, Arnold	Hip-joint replacement	2
Bernard, Stephen	Cape Cod Potato Chips	*3*, *5*, 7, *8*
Bianchi, Al	Lotus cars	*9*
Bic	*Disposable razors*	*9*
Biomagnetic Tech.	*Magnetic Source Imaging*	*3*
Black, James	*Tagamet*	*1*
BMW	*Rear side window design and Acura*	*5*
Boyer, Herbert	*Recombinant DNA*	*11*
Boulnois, Jean-Luc	President, Technomed U.S.	*7*
Bristol-Myers	Antiperspirant technology	11
Bristol-Myers Squibb	Windex	5
Carlson, Chester	Xerox, Haloid Corp., Battelle Institute	3
Carothers, William	Nylon	1
Carpino, Joe	Listerine	11
Carter-Wallace	Arrid	5
Causey, Zane	New product showcase	*8*
Chakrabarty, Ananda	Recombinant DNA	11
Chavkin, Len	Aerosol whip	4
Clemens, George	Interplak	6, *8*, 14
Clorox Co.	Clorox Super Detergent	8
Coca-Cola Co.	Value of good brand name	5
	Formula: trade secret	5
	New Coke	8
Cohen, Stanley	Recombinant DNA	*11*
Colorocs	Copying machine	1
Colt, Samuel	Mass production—interchangeable parts	1
Cot'n Wash	Trademark	5

Name/Brand	*Invention/Subject*	*Chapter(s)*
Crum, George	Saratoga Chips	5
Cuervo, Armando	SleepTight, SweetDreams, Inc.	3, *8*
Cuisinart	Food processor	1
Damadian, Raymond	MRI	*4*
Dean, Robert	Designer Blocks, Landscape Blocks	*12*, 13
DeLeon, Richard	Toothpaste/toothbrush bathroom wall fixture	5
Diamond Fever Gang	Synthetic diamonds	6
Dichter, Ernest	Tea market research	8
Dowd, Fred J.	Folding corrugated box	*12*
Du Pont	Nylon trademark	5
	Lanxide joint venture	3
	$EED program	14
Dynagran Corp.	Marion Laboratories and calcium supplement	5
Eastman, George	First cameras with roll film	1
Eastman Kodak	Instant photography	4
Edison, Thomas	Light bulb and invention lab	1, 2
Efferdent	Denture cleaner	1
Eisenberg, Joe	Tofutti	9
Elan Pharmaceut.	Licensing technology	10
Endless Pools	Swimming machine	1, 3
Engelberg, Al	Patent attorney	4
Everlast	Trademark	5
Explore Technology	Video transmittal	13
Fleming, Alexander	Penicillin	13
Fonar Corp.	MRI	4
Ford, Henry	Spark plug	1
Fraden, Jacob	Infrared thermometer	13
Franklin Mint	The made-up name	5
Frase, E. C.	Attached can pull-ring Dayton Reliable, Alcoa	9
Friel, Dan	Chef's Choice	2, *4*, 7, 8, 9, 10, *12*
Frito-Lay	Potato chip wars	5, 9
Fry, Art	Post-it Notes	1, 2, *9*, 10
Geneen, Harold	Intrapreneurship	14
Genentech	Human insulin, Eli Lilly	11

Name/Brand	*Invention/Subject*	*Chapter(s)*
General Electric	Diamond Fever Gang	14
General Foods	Market research method	8
General Motors	Saturn automobile	9
Genpharm, Inc.	Glaxo Holdings and Zantac	4
Gershon, Sol	Clear gel toothpaste	9
Gillette	Wilkinson Sword Stainless Blade	9
	Fighting off takeover bids	9
	On outside invention	10
Goldscheider, Bob	International Licensing Network, plastic link chain and contracts	10
Goodrich, Marshall	Fluoride	1
Goodyear, Charles	Loss of rights and cases	4
Gorman's New Product News	Not-so-new products	9
Gould, Gordon	Lasers	*3*, *6*, 11
Gray, Elisha	Telephone	1
Greatbatch, Wilson	Pacemaker	1
Green, Barrett	Carbonless carbon paper	2, *10*
Green, Herbert	Denture solution invention	1
Greene, William	Tennis ball machine	3
Harris, Art	Metering valve for aerosols	9, 12
Hascöet, Gérard	Technomed	7
Heidelberger, Lou	Techlex Division of Reed, Smith, Shaw and McClay	4
Hershey	Trade secrets	5
Hickman, Ron	WorkMate	*12*, 13
Hidden Beauty	Natural trademark	5
Hiles, Maurice	Sorbothane	2, *10*
Hodock, Cal	Marketing expert	8
Hogan, C. Lester	Motorola and Fairchild Camera and trade secrets	5
Hood Dairy	Lichee nut ice cream	8
Hoyt, Harry	Carter-Wallace, agreements	11

Name/Brand	*Invention/Subject*	*Chapter(s)*
Interplak	Electric toothbrush	6, *8*
Inventor organizations	Value of	11
Ishler, Norman	Aerosol effervescent	2, *10*
James, David	Re-Tech dry battery technology	*11*
Jamison, Tony	Nicholas-Kiwi	7
Johnson & Johnson	Damadian lawsuit, Sneider lawsuit, 3M	4
Kahn, Bernie	Pricing ideas	11
Kearns, Robert	Intermittent windshield wiper	*4*
Kiam, Victor	Remington Electric Shaver	9
Kimberly-Clark	Sneider lawsuit, Lightdays	4
Kreible, Vernon	Loctite	*9*
Kurzweil, Raymond	Electronic scanner for blind	1
Lang, Gene	Refac	3, 4
Lanxide Corp.	Aluminum/ceramic materials	3, 4, *10*, *13*
Latimer, Lewis	Worked with A. G. Bell	4
Lay, Herman	Lay's Potato Chips	5
Lawrence, J. J.	Listerine formula	11
Lemelson, Jerry	Mattel Hot Wheels	4
Levine, Ralph	Carter-Wallace	11
Levi's	History of a brand	5
Lindseth, Jonathan	Interplak	6, 8
Lindseth, Steven	Thermoscan	13
Listerine	Licensing	*11*
Loafers	Filling a need	8
Loctite	Adhesives	*9*
Lyden, Michael	Tyco toys	4, 7
Mackles, Len	Aerosol Whip	4
	Liquid powder technology	11
Main Street Toy Company	Slap Wrap	5
Marra, Dorothy	Aerosol pad	2
Martin, Andrew	Smartfood popcorn	9
Masco Corp.	Outreach program	14
Mattel Corp.	Hot Wheels	4
McMath, Robert	The Marketing Showplace	8

Name/Brand	*Invention/Subject*	*Chapter(s)*
	Single-use paint container with applicator	9
Merck & Co.	Theft of drug formulations	5
Minnesota Mining (3M)	Cast tape trade secret	5
	Internal invention program	14
Montague, David and Harry	BiFrame bike	2, 7, *12*, 13
Montgomery Securities	Venture capital and start-ups	2
Motorola	Trade secrets	5
Muench, Charles	Colorocs	1, *2*, 3, 10
Murdock, James	Endless Pools	1, 3
Murtha, Richard	Slap Wrap, Main Street Toy Company	*5*
Myers, Ken	Smartfood popcorn	9
National Analysts	Lifestyle segmentation	8
Nelson, John	BiFrame bike	12
Nesmith, Betty	Typewriter correction fluid	12
Newkirk, Marc	Lanxide Corp.	*3*, *4*, 10, 13
Nintendo	Suits against Atari and Lewis Galoob	5
Norfolk v. Peabody	1868 case	5
North American Philips	Shavers	4
Nova Pharmaceutical Corporation	Bradykinin-inhibitors	13
Nylon	Trademark lapse	5
Oral Research, Inc.	Marketer of PLAX	1
Onanian, Rick	Learning Things, Inc.	*12*
Osipow, Lloyd	Aerosol pad	2
Otis, E. G.	Elevator	1
Pader, Morton	Close-up toothpaste	9
Panoz, Don	Elan Pharmaceuticals PLC	10
Pepsico	Pepsi brand name	5
Perrier	Benzene	5
Pfizer, Inc.	Japanese drug license, Thornton-SuperFloss	11
Pharmacraft Corp.	Foot bath	10
Phipard, Henry	Multilobular screws	5, 6, *10*
Picturesque Stockings	Design patent suit	5

Index